Studies in Space Policy

Volume 28

Series Editor

European Space Policy Institute, Vienna, Austria

The use of outer space is of growing strategic and technological relevance. The development of robotic exploration to distant planets and bodies across the solar system, as well as pioneering human space exploration in earth orbit and of the moon, paved the way for ambitious long-term space exploration. Today, space exploration goes far beyond a merely technological endeavour, as its further development will have a tremendous social, cultural and economic impact. Space activities are entering an era in which contributions of the humanities — history, philosophy, anthropology —, the arts, and the social sciences — political science, economics, law — will become crucial for the future of space exploration. Space policy thus will gain in visibility and relevance. The series Studies in Space Policy shall become the European reference compilation edited by the leading institute in the field, the European Space Policy Institute. It will contain both monographs and collections dealing with their subjects in a transdisciplinary way.
The volumes of the series are single-blind peer-reviewed.

More information about this series at http://www.springer.com/series/8167

Annette Froehlich · Nicolas Ringas ·
James Wilson

Space Supporting Africa

Volume 3: Security, Peace, and Development
through Efficient Governance Supported
by Space Applications

Annette Froehlich (iD)
European Space Policy Institute
Vienna, Austria

Nicolas Ringas (iD)
European Space Policy Institute
Vienna, Austria

James Wilson (iD)
European Space Policy Institute
Vienna, Austria

ISSN 1868-5307 ISSN 1868-5315 (electronic)
Studies in Space Policy
ISBN 978-3-030-52262-9 ISBN 978-3-030-52260-5 (eBook)
https://doi.org/10.1007/978-3-030-52260-5

This Springer imprint is published by the registered company Springer Nature Switzerland AG
The registered company address is: Gewerbestrasse 11, 6330 Cham, Switzerland

Preface

This volume considers the leveraging of space applications to enable efficient governance in African civil society, thereby supporting security, peace and development within the continent. It analyses the potential benefits of e-governance for African countries and describes how space-based applications can be employed in implementing e-government solutions, thereby strengthening civil society, bolstering democratic processes and supporting socio-economic development, in line with Africa's development goals. As such, this volume reviews the landscape of governance and e-governance both globally and within the African context and then identifies the most fruitful avenues for space technologies and applications to support the strengthening of African governance.

Chapter 1 explores the positive correlation between good governance and achieving the United Nations (UN) Agenda 2030 Sustainable Development Goals (SDGs) and the African Union (AU) Agenda 2063 aspirations. Subsequently, the various existing governance structures and frameworks established at a continental level within Africa are analysed, namely: the African Union (AU), African policy frameworks relevant to governance, the New Partnership for Africa's Development (NEPAD), the African Peer Review Mechanism (APRM) and the African Continental Free Trade Area (AfCFTA). Specific challenges threatening governance within Africa are examined to determine remedial actions that can be employed to overcome them. The current governance levels in Africa are assessed by examining the results of the latest African Peer Review Mechanism's African Governance Report and the Ibrahim Index of African Governance study, and areas where improvements are necessary are identified.

Chapter 2 assesses e-government readiness levels across Africa through an analysis of the key findings from the United Nations E-Government Survey from 2018. Thereafter, new trends and emerging technologies in e-government including blockchain, e-voting and digital government are discussed. Specific case studies of innovative e-government applications from both developed and developing countries are examined, to identify objectives and strategies that African countries should consider when implementing their own e-government solutions.

Chapter 3 investigates how the digital divide and lack of Internet connectivity within Africa can be overcome using new technologies and commitments from both terrestrial and space actors. The opportunities facilitated by rapidly growing satellite constellations that increase connectivity levels and reduce costs are analysed. The various planned constellations are assessed on a technical level.

Chapter 4 examines space-based applications that can be utilised to improve governance and support e-governance in different sectors, specifically: (i) peace, security and defence, (ii) healthcare and disease control, (iii) water provision and management, (iv) biodiversity management. The chapter identifies specific sources for African countries to obtain Earth Observation and remote sensing data to minimise costs and maximise benefits. Various geographical information system (GIS) software packages used to process and manipulate remote sensing data are discussed, after which relevant European Space Agency projects focused on supporting African development are analysed, including Copernicus, SWAY4Edu2, GlobWetland Africa, B-LiFE, SatFinAfrica and the ESA Tiger project. Lastly, Chapter 4 explores how space-based applications can support the UN SDGs by modelling the risks, interactions and costs of the various SDGs.

Vienna, Austria   Annette Froehlich
                                                                                               Nicolas Ringas
                                                                                                James Wilson

Contents

About the Authors

Dr. Annette Froehlich is a scientific expert seconded from the German Aerospace Centre (DLR) to the European Space Policy Institute (Vienna), and an honorary adjunct senior lecturer at the University of Cape Town (SA) at SpaceLab. She graduated in European and International Law at the University of Strasbourg (France), followed by business-oriented postgraduate studies and her Ph.D. at the University of Vienna (Austria). Responsible for DLR and German representation to the United Nations and International Organisations, she was also a member/alternate head of delegation of the German delegation to UNCOPUOS. Moreover, Dr. Annette Froehlich is an author of a multitude of specialist publications and serves as a lecturer at various universities worldwide in space policy, law and society aspects. Her main areas of scientific interest are European space policy, international and regional space law, emerging space countries, space security and space & culture. She has also launched, as editor, the new scientific series "Southern Space Studies" (Springer publishing house) dedicated to Latin America and Africa. e-mail: Annette.Froehlich@espi.or.at; Annette.Froehlich@dlr.de

Nicolas Ringas graduated with a Bachelor of Science degree in electrical engineering from the University of the Witwatersrand, in Johannesburg, South Africa, in 2012. Since graduating, Nicolas has been working at an engineering consultancy firm in the water, oil and natural gas sector, specialising in electromagnetic interference issues with AC power lines, railways and cables. He is currently completing a Master's of Philosophy in space sciences at the University of Cape Town in South Africa. e-mail: nicringas@gmail.com

James Wilson graduated from electrical engineering at the University of the Witwatersrand in 2017. He has a keen interest in space technology and hence pursued a Master of Philosophy in space science at the University of Cape Town in 2018. He successfully completed his dissertation in designing a control system for a liquid rocket test stand. He is currently working on renewable energy projects throughout Africa, especially the design and testing of large solar and wind farms. e-mail: Jameswilson9@icloud.com

The authors specially acknowledge the assistance of **André Siebrits**, a South African researcher and Associate Editor of Southern Space Studies (Springer Series) for his advice and exchange of publication experience. Mr. Siebrits focuses on space and international relations, especially in developing world contexts, as well as on education and the use of educational technologies. e-mail: andre.siebrits@southernspacestudies.com

Abbreviations

ACBF	African Capacity Building Foundation
ACDEG	African Charter on Democracy, Elections and Governance
ACERWC	African Committee of Experts on the Rights and Welfare of the Child
ACHPR	African Union Commission on Human and Peoples' Rights
AEC	African Economic Community
AfCFTA	African Continental Free Trade Area
AfCHPR	African Court on Human and Peoples' Rights
AfDB	African Development Bank
AGA	African Governance Architecture Platform
AGR	African Governance Report
AI	Artificial Intelligence
AIDS	Acquired Immune Deficiency Syndrome
AISI	African Information Society Initiative
AML	Anti-Money Laundering
APAI-CRVS	Africa Programme on Accelerated Civil Registration and Vital Statistics
APB	Aadhaar Payment Bridge
APRM	African Peer Review Mechanism
APSA	African Peace and Security Architecture
ARFSD	African Regional Forum on Sustainable Development
ATM	Automated Teller Machine
AU	African Union
AUABC	African Union Advisory Board on Corruption
AUC	African Union Commission
AUCIL	African Union Commission on International Law
AV-net	Anonymous Veto Protocol
CAA	Civil Aviation Authority
CAADP	Comprehensive Africa Agriculture Development Programme
CERT	Computer Emergency Response Team
CERT-MU	Computer Emergency Response Team of Mauritius

CFTA	Continental Free Trade Area
CIDR	Central Identities Data Repository
CII	Critical Information Infrastructure
CIMER	Turkey's Presidency's Communication Centre
CNES	Centre National D'Etudes Spatiales
COP	Child Online Protection
CPI	Corruption Perception Index
CRVS	Civil Registration and Vital Statistics
CSE	Communications Security Establishment (Canada)
CSIRT	Computer Security Incident Response Team
CSSDCA	Conference on Security, Stability, Development and Co-operation in Africa
DARPA	Defense Advanced Research Projects Agency
DBT	Direct Benefit Transfer
DEM	Digital Elevation Map
DFID	UK Department for International Development
DLT	Distributed Ledger Technology
DRC	Democratic Republic of Congo
DRE	Direct Recording Electronic voting system
EDRS	European Data Relay System
EGDI	E-Government Development Index
EIU	Economist Intelligence Unit
EO	Earth Observation
EPI	E-Participation Index
ESA	European Space Agency
EU	European Union
EVM	Electronic Voting Machine
FDA	US Food and Drug Administration
FTYIP	First Ten-Year Implementation Plan
GCA	Global Cybersecurity Agenda
GCI	Global Cybersecurity Index
GDP	Gross Domestic Product
GDPR	General Data Protection Regulation
GEO	Geostationary Orbit
GIFMIS	Ghana Integrated Financial Management Information System
GIS	Geographic Information System
GNSS	Global Navigation Satellite System
GoG	Government of Ghana
GPS	Global Positioning System
GSM	Global System for Mobile Communications
HCI	Human Capital Index
HHS	Health and Human Services
HLPF	High-Level Political Forum on Sustainable Development
HSS	US Department of Health and Human Services
IAEG-SDG	Inter-agency and Expert Group on SDG Indicators

ICT	Information and Communications Technology
ID4D	Identification for Development
IDA	International Development Association
IEG	Independent Evaluation Group
IFZ	Institute for Financial Services (Lucerne University)
IIAG	Ibrahim Index of African Governance
IoT	Internet of Things
IPRA	International Public Relations Association
ISS	International Space Station
ITES	Information Technology Enabled Services
ITU	International Telecommunications Union
JRC	Joint Research Centre
KYC	Know Your Customer
LEO	Low Earth Orbit
LPG	Liquefied Petroleum Gas
MAP	Millennium Africa Recovery Plan
MGNREGA	Mahatma Gandhi National Rural Employment Guarantee Act
MIF	Mo Ibrahim Foundation
MODIS	Moderate Resolution Imaging Spectroradiometer
MOSFET	Metal Oxide Semiconductor Field Effect Transistor
NASA	National Aeronautics and Space Administration
NDVI	Normalised Difference Vegetation Index
NEC	National Electoral Commission
NEPAD	New Partnership for Africa's Development
NGO	Non-Governmental Organisation
NHS	National Health Service
NIR	Near-Infrared
NPCI	National Payments Corporation of India
NSO	National Statistical Office
OAU	Organisation of African Unity
ODIN	Open Data Inventory
OECD	Organisation for Economic Co-operation and Development
OGD	Open Government Data
OPSI	Observatory of Public Sector Innovation
OSAA	Office of the Special Advisor on Africa
OSI	Online Services Index
OTP	One Time Pin
OVN	Open Vote Network
PARIS21	Partnership in Statistics for Development in the 21st Century
PDS	Public Distribution System
PHC	Population and Housing Census
PMJDY	Pradhan Mantri Jan Dhan Yojana
POS	Point-of-Sale
PPP	Public–Private Partnership
PRC	Permanent Representatives Committee

REC	Regional Economic Communities
RGB	Red Green Blue
RSS	Rich Site Summary
SAATM	Single African Air-Transport Market
SAIIA	South African Institute for International Affairs
SAR	Synthetic Aperture Radar
SDG	Sustainable Development Goals
SHaSA	Strategy for the Harmonisation of Statistics in Africa
SIM	Subscriber Identification Module
SMOS	Soil Moisture and Ocean Salinity
SMS	Short Message Service
SSA	Sub-Saharan Africa
STATAFRIC	African Union Institute for Statistics in Africa
STC	Specialised Technical Committees
TII	Telecommunications Infrastructure Index
TRIPS	Total Revenue Integrated Processing System
TYIP	Ten-Year Implementation Plan
UAS	Unmanned and Autonomous System
UHF	Ultra High Frequency
UID	Unique Identification Number
UIDAI	Unique Identification Authority of India
UN	United Nations
UN DESA	United Nations Department of Economic and Social Affairs
UNDP	United Nations Development Programme
UNECA	United Nations Economic Commission for Africa
UNESCO	United Nations Educational, Scientific and Cultural Organisation
UNFPA	United Nations Population Fund
UNHCR	United Nations High Commissioner for Refugees
UNICEF	United Nations Children's Fund
UNSD	United Nations Statistics Division
USA	United States of America
UVI	Unique Verifiable Identity
VHF	Very High Frequency
VNR	Voluntary National Review
VSAT	Very Small Aperture Terminal
WAN	Wide Area Network
WHO	World Health Organisation
WSIS	World Summit on the Information Society

Chapter 1
A Look at Governance Throughout Africa

Abstract This chapter defines governance and explores how good governance is integral to realizing the United Nations Agenda 2030 Sustainable Development Goals and the African Union Agenda 2063 aspirations. It then examines the various governance structures established within Africa in order to provide an overview of the frameworks that inform governance within the continent, as well as the specific challenges that need to be overcome to ensure good governance. An analysis of the current governance levels in Africa is presented through an examination of the African Peer Review Mechanism's African Governance Report and the Ibrahim Index of African Governance study and critical areas requiring improvements are identified.

1.1 Introduction

> Development is impossible in the absence of true democracy, respect for human rights, peace and good governance[1]

New Partnership for Africa's Development

The United Nations Children's Fund (UNICEF) forecasts that by 2050, more than 2 billion babies will be born in Africa. Similarly, the Word Economic Forum estimates that by 2055 there will be approximately 1.5 billion Africans aged between 15 and 64 years.[2] The growth rate of this age group is known as Africa's demographic dividend and yields an opportunity for economic development.

The UN estimates that by 2050, Africa will account for more than one quarter of the world's projected working age population. Moreover, it is estimated that the

---

[1] African Peer Review Mechanism, *Strategic Plan 2016–2020*, South Africa, 2016.

[2] J. McKenna, *Six numbers that prove the future is African*, World Economic Forum, May 2017, www.weforum.org/agenda/2017/05/africa-is-rising-and-here-are-the-numbers-to-prove-it/, accessed: 8 February 2020.

A. Froehlich et al., *Space Supporting Africa*, Studies in Space Policy 28, https://doi.org/10.1007/978-3-030-52260-5_1

African economy will exhibit a growth rate double that of the developed world.[3] Good governance throughout Africa is paramount to realizing the socio-economic opportunities afforded by its human capital.

In the 2008 Compendium of Basic UN Terminology (UNTERM) the definition for "governance" is:

> The exercise of political, economic and administrative authority in the management of a country's affairs at all levels. Governance is a neutral concept referring to the complex mechanisms, processes, relationships and institutions through which citizens and groups articulate their interests, exercise their rights and obligations and mediate their differences.[4]

Similarly, the definition of the Office of the UN High Commissioner for Human Rights states:

> Governance is the process whereby public institutions conduct public affairs, manage public resources and guarantee the realization of human rights. Good governance accomplishes this in a manner essentially free of abuse and corruption, and with due regard for the rule of law. The true test of *good* governance is the degree to which it delivers on the promise of human rights: civil, cultural, economic, political and social rights.[5]

This chapter examines the existing governance levels in African states, investigates the challenges facing good governance within Africa and identifies fundamental areas where improvements are required. The second chapter explores how e-governance and e-government applications can be employed to assist with realizing good governance throughout the continent. In this regard, the European Union (EU) defines e-government as "the use of information and communication technology in public administrations combined with organisational change and new skills in order to improve public services and democratic processes and strengthen support to public policies".[6]

1.2   Africa's Development Goals

Prior to examining governance throughout Africa, it is first necessary to discuss the development goals of African states, which are influenced by the UN Agenda 2030 and the AU Agenda 2063. Good governance is crucial to realize the two agendas and numerous countries have revised their national development strategies to align with these goals and aspirations. Furthermore, tracking the progress achieved towards the two agendas offers insights into governance efficiency in African states.

---

[3]United Nations Economic Commission for Africa, *African Continental Free Trade Area—Questions & Answers*, United Nations Economic Commission for Africa, www.uneca.org/publications/african-continental-free-trade-area-questions-answers, accessed: 27 April 2020.

[4]United Nations Economic and Social Council, *Compendium of basic United Nations terminology in governance and public administration*, New York, 2008 (E/C.16/2008/3), p. 23.

[5]Ibid., p. 24.

[6]Ibid., p. 25.

1.2.1 African Union Agenda 2063

Agenda 2063[7] is a 50 year framework seeking to "prioritize inclusive social and economic development, continental and regional integration, democratic governance and peace and security amongst other issues aimed at repositioning Africa to become a dominant player in the global arena." It forms a strategic blueprint that details how Africa can transform itself over the 50 years between 2013 and 2063 to attain the Pan-African vision of "an integrated, prosperous and peaceful Africa, driven by its own citizens, representing a dynamic force in the international arena."[8]

Agenda 2063, developed under the auspices of the AU Commission (AUC), was formally adopted in 2015 and comprises seven aspirations each with their own goals (see Table 1.1). It aspires towards "shared prosperity and well-being, for unity and integration, for a continent of free citizens and expanded horizons, where the full potential of women and youth are realized, and with freedom from fear, disease and want".[9]

Agenda 2063 identifies key activities in five separate Ten Year Implementation Plans (TYIP), which provide both quantitative and qualitative metrics to assess and monitor progress in the form of goals and targets. The purpose of the First Ten-Year Implementation Plan (FTYIP) is to:

- "Identify priority areas, set specific targets, define strategies and policy measures to implement the FTYIP of Agenda 2063.
- Bring to fruition the Fast Track programmes and initiatives outlined in the Malabo Decisions[10] of the AU to provide the big push and breakthroughs for Africa's economic and social transformation.
- Provide information to all key stakeholders at the national, regional and continental levels on the expected results/outcomes for the first ten years of the plan

---

[7]For a further discussion of Agenda 2063 in the context of the African space sector, see: A. Froehlich and A. Siebrits, *Space Supporting Africa Volume 1: A Primary Needs Approach and Africa's Emerging Space Middle Powers* (Cham: Springer, 2019).

[8]The African Union Commission, *Agenda 2063: The Africa We Want*, https://au.int/en/agenda2063/overview, accessed: 21 January 2020.

[9]The African Union Commission, *Our Aspirations for the Africa We Want*, https://au.int/en/agenda 2063/aspirations, accessed: 21 January 2020.

[10]At the AU Summit of June 2014 in Malabo (Equatorial Guinea) the AU Assembly identified high priority programmes to highlight the success of Agenda 2063 to the African Citizenry. The projects and programmes identified were: the Integrated High Speed Train Network, increasing efforts towards establishing a Continental Free Trade Area, the African Passport and free movement of people, establishing the Single African Aviation Market, establishing various financial institutions at continental level (such as the African Investment Bank, the African Remittances Institute, the African Credit Guarantee Facility, the African Monetary Union and the African Central Bank), implementing the Grand Inga Dam Project hydroelectric power station, setting up the Pan African E-Network to allow for increased infrastructure and cybersecurity to support e-governance, creating the Annual Consultative Platform to assist in reaching the Agenda 2063 Aspirations, and the Silencing the Guns by 2020 initiative. These high priority programmes were chosen to be fast tracked and as such were included in the FTYIP.

Table 1.1 AU Agenda 2063 aspirations and goals

Aspiration	Description	Goals
Aspiration 1	A prosperous Africa based on inclusive growth and sustainable development	1. A high standard of living, quality of life and well-being for all
		2. Well educated citizens and skills revolutions underpinned by science, technology and innovation
		3. Healthy and well-nourished citizens
		4. Transformed economies and jobs
		5. Modern agriculture for increased proactivity and production
		6. Blue/Ocean Economy for accelerated economic growth
		7. Environmentally sustainable climate and resilient economies and communities
Aspiration 2	An integrated continent, politically united and based on the ideals of Pan-Africanism and the vision of Africa's Renaissance	1. Establishing a United Africa (Federal/Confederate) and realizing a Continental Free Trade Area (CFTA)
		2. World class infrastructure criss-crosses Africa
		3. Decolonisation
Aspiration 3	An Africa of good governance, democracy, respect for human rights, justice and the rule of law	1. Democratic values, practices, universal principles for human rights, justice and rule of law entrenched
		2. Capable institutions and transformed leadership in place at all levels

(continued)

and assign responsibilities to all stakeholders in its implementation, monitoring and evaluation.

- Outline the strategies required to ensure availability of resources and capacities together with citizen's engagement in the implementation of the First Ten Year Plan."[11]

The FTYIP covers the period from 2013 to 2023 and identifies 20 Agenda 2063 goals linked to the seven aspirations (as shown in Table 1.1).[12] Furthermore, priority areas for national level implementation are identified for each of the goals, along with their respective targets. In total there are 39 priority areas and 255 targets, as summarised in Fig. 1.1. Of those targets, 63 are classified as core indicators and the

---

[11] The African Union Commission, *The First-Ten Year Implementation Plan*, https://au.int/en/age nda2063/ftyip, accessed: 21 January 2020.

[12] Ibid.

Table 1.1   (continued)

Aspiration	Description	Goals
Aspiration 4	A peaceful and secure Africa	1. Peace, security and stability is preserved
		2. A stable and peaceful Africa
		3. A fully functional and operational African Peace and Security Architecture (APSA)
Aspiration 5	An Africa with a strong cultural identity, common heritage, shared values and ethics	1. Africa cultural renaissance is pre-eminent—inculcating the spirit of Pan Africanism, ensuring creative arts are major contributors to Africa's growth, and restoring and preserving African cultural heritage
Aspiration 6	An Africa, whose development is people-driven, relying on the potential of African people, especially its women[a] and youth, and caring for children	1. Full gender equality in all spheres of life
		2. Engaged and empowered youth and children
Aspiration 7	Africa as a strong, united, resilient and influential global player and partner	1. Africa as a major partner in global affairs and peaceful co-existence
		2. Africa takes full responsibility for financing her development

[a] For an overview of the perspectives of African women on the African space sector, see T. Oniosun, N. M. A. Ndieguene, M. Mwamba, et al., "African Women Competition", in *Space Fostering African Societies: Developing the African Continent through Space, Part 1*, ed. Annette Froehlich (Cham: Springer, 2020), 253–272

The African Union Commission, *Our Aspirations for the Africa We Want*, https://au.int/en/agenda 2063/aspirations, accessed: 21 January 2020

FTYIP requires all states to report on these indicators to their Regional Economic Communities (RECs) as a minimum requirement.[13]

The objectives and key focus areas of the FTYIP were influenced by the National Development Plans of member states and the strategic plans of the RECs, existing continental frameworks, the key Transformational Outcomes of Agenda 2063 and, lastly, by Flagship Projects identified within Agenda 2063.[14]

The Flagship Projects are 13 key initiatives encompassing infrastructure, education,[15] science, technology, arts and culture and are aimed at boosting the growth and

---

[13] Mo Ibrahim Foundation, *African Governance Report—Agendas 2063 & 2030: Is Africa on Track?*, 2019.

[14] The African Union Commission, *The First-Ten Year Implementation Plan*, https://au.int/en/age nda2063/ftyip, accessed: 21 January 2020.

[15] For additional information regarding education in Africa in the context of the space sector, see: A. Siebrits and V. van de Heyde, "Towards the Sustainable Development Goals in Africa: The African Space-Education Ecosystem for Sustainability and the Role of Educational Technologies", in *Embedding Space in African Society: The United Nations Sustainable Development Goals 2030*

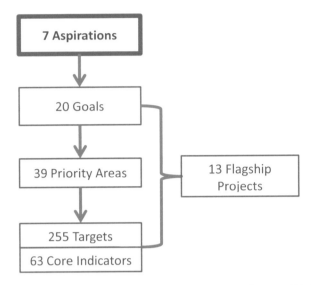

Fig. 1.1 Overview of the AU Agenda 2063 goals, priority areas and targets (Mo Ibrahim Foundation, *African Governance Report—Agendas 2063 & 2030: Is Africa on Track?*, 2019, Sect. 1.1, p. 11)

development of the African economy.[16] The Flagship Projects identified by Agenda 2063 are listed below[17]:

1. Integrated high-speed train network
2. African commodity strategy
3. An African Continental Free Trade Area (AfCFTA)
4. African passport and freedom of movement
5. Silencing the Guns by 2020
6. Implementation of the Grand Inga Dam Project
7. Establishing a Single African Air-Transport Market (SAATM)
8. Establishing an Annual African Economic Forum
9. Establishing African financial institutions including an African Investment Bank, a Pan African Stock Exchange, the African Monetary Fund and the African Central Bank
10. Pan-African e-Network

Supported by Space Applications, ed. A. Froehlich (Cham: Springer, 2019), 127–180; C. Müller, "Aerospace Research in African Higher Education", in *Embedding Space in African Society: The United Nations Sustainable Development Goals 2030 Supported by Space Applications*, ed. A. Froehlich (Cham: Springer, 2019), 113–126; and A. Durczok, "Time to Change Your Education Programme—The Transformative Power of Digital Education", in *Space Fostering Latin American Societies: Developing the Latin American Continent trough Space, Part 1*, ed. A. Froehlich (Cham: Springer, 2020), 115–124.

[16]The African Union Commission, *Flagship Projects of Agenda 2063*, https://au.int/agenda2063/flagship-projects, accessed: 21 January 2020.

[17]Ibid.

11. Africa outer space strategy
12. An African virtual and e-university
13. Cyber security programme guided by the AU Convention on Cyber Security and Personal Data Protection.

It is noteworthy that African activities and involvement in space activities are addressed as one of the 13 specific programmes listed in Agenda 2063.[18] Agenda 2063 identifies the importance and potential of space-related activities and states that "outer space is of critical importance to the development of Africa in all fields: agriculture, disaster management, remote sensing, climate forecast, banking and finance and defense and security." Furthermore, two other central projects, namely the Pan-African e-Network and the African e-University intend to capitalize on technological breakthroughs in Internet connectivity to transform Africa into an e-society complete with distance learning, e-learning and e-applications.[19] Successful implementation of all 13 Flagship Projects will require extensive use of space-based technologies.

1.2.2 United Nations Agenda 2030 Sustainable Development Goals

In September 2015, the UN General Assembly adopted Agenda 2030 on Sustainable Development,[20] which "provides a shared blueprint for peace and prosperity for people and the planet."[21] The Agenda consists of four parts, namely: (i) vision and principles, (ii) goals and targets, (iii) means of implementation and (iv) follow up and review mechanisms.[22]

In total, Agenda 2030 has 17 sustainable development goals (SDGs) (as depicted in Fig. 1.2), consisting of 169 targets and 231 indicators. The targets described in the Agenda are on a global, aspirational level and are intended as a guide for each state in implementing their own national targets.[23]

---

[18]For an in-depth selection of African country profiles pertaining to the space sector, see: A. Froehlich (ed.), *Integrated Space for African Society: Legal and Policy Implementation of Space in African Countries* (Cham: Springer, 2019).

[19]Ibid.

[20]For more information on the Sustainable Development Goals in the context of the African space sector, see: A. Froehlich (ed.), *Embedding Space in African Society: The United Nations Sustainable Development Goals 2030 Supported by Space Applications* (Cham: Springer, 2019); S. Wade (ed.), *Earth Observations and Geospatial Science in Service of Sustainable Development Goals: 12th International Conference of the African Association of Remote Sensing of the Environment* (Cham: Springer, 2019); A. Siebrits, O. Mookeletsi, A. Alberts, and A. Gairiseb, "Africa and Space", in *Integrated Space for African Society: Legal and Policy Implementation of Space in African Countries*, ed. A. Froehlich (Cham: Springer, 2019), 27–55; and V. Munsami, "Maximising the Use of Space Applications in Implementing the Sustainable Development Goals in Africa", in *Embedding Space in African Society: The United Nations Sustainable Development Goals 2030 Supported by Space Applications*, ed. A. Froehlich (Cham: Springer, 2019), 1–30.

[21]United Nations, Sustainable Development Goals Knowledge Platform, Sustainable Development Goals, https://sustainabledevelopment.un.org/sdgs, accessed: 21 January 2020.

[22]United Nations South Africa, 2030 Agenda, www.un.org.za/sdgs/2030-agenda/, accessed: 21 January 2020.

[23]Ibid.

Fig. 1.2 United Nation's Agenda 2030 Sustainable Development Goals (SDGs) (United Nations, Sustainable Development Goals Knowledge Platform, Sustainable Development Goals, https://sustainabledevelopment.un.org/sdgs, accessed: 21 January 2020)

The High-Level Political Forum on Sustainable Development (HLPF) is mandated with monitoring progress towards the SDGs on an annual basis. This process is completed on a four year cycle, ensuring every goal is reviewed once every four years. In order to assist progress review sessions, the UN Department of Economic Social Affairs (UN DESA) issued a Voluntary National Review (VNR) handbook that suggests a framework for each country to perform a voluntary self-assessment on its progress.[24] The monitoring process was enhanced in 2015 when the UN Statistical Commission created the Inter-agency and Expert Group on SDG Indicators (IAEG-SDGs), which developed an online tracking tool where progress indicators can be accessed, provided data is available.[25]

The African Regional Forum on Sustainable Development (ARFSD) is a multi-stakeholder platform that monitors progress on the SDGs and Agenda 2063 on an annual basis by assessing SDGs and their aligned Agenda 2063 targets due for HLPF

---

[24] UN Department of Economic and Social Affairs, High-Level Political Forum on Sustainable Development, *Handbook for the preparation of voluntary data reviews*, 2019.

[25] United Nations Department of Economic and Social Affairs, Statistics Division, *SDG Monitoring and Reporting Toolkit for U Country Teams*, https://unstats.un.org/sdgs/unct-toolkit/, accessed: 21 January 2020 (Individual country profiles available at: https://country-profiles.unstatshub.org/).

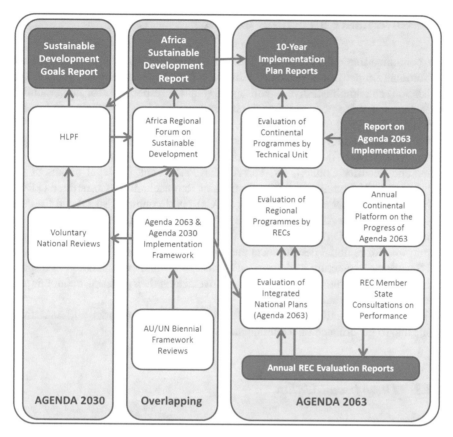

Fig. 1.3 Monitoring Agenda 2030 and Agenda 2063—important overlaps and reporting structures (Mo Ibrahim Foundation, *African Governance Report*, 2019, Sect. 1.1, p. 15)

review in the following year. To facilitate the review process and avoid duplication, the AUC and the UN signed the *AU-UN Framework for the Implementation of Agenda 2063 and the Agenda 2030 for Sustainable Development*, which suggests common reporting structures and assessments as well as biennial reviews of the implementation process.

The overlapping nature of monitoring both Agendas has, however, received some criticism, with certain actors stating that the process is overcomplicated and shared accountability between various stakeholders is not accurately quantified.[26] Figure 1.3 highlights the different structures involved in the monitoring processes of the two Agendas and identifies overlaps in the processes.

---

[26]Mo Ibrahim Foundation, *African Governance Report*, 2019.

1.3   Governance Structures in Africa

The continued implementation and functioning of governance require an elaborate institutional framework of political institutions (that draft and enforce policies and legislation) and administrative organisations (that are responsible for implementing and upholding those policies), functioning across three levels, namely continental, regional and national (or member state).

On a continental level, the AU and its respective organs are of paramount importance, specifically: (i) the Assembly of the Union, (ii) Executive Council, (iii) Peace and Security Council, (iv) Pan-African Parliament, (v) legal organs of the AU, (vi) the AU Commission, (vii) Permanent Representatives Committee (PRC), (viii) Specialised Technical Committees (STCs), (ix) Economic, Social and Cultural Council, and (x) The African Committee of Experts on the Rights and Welfare of the Child (ACERWC). On a regional level, the AU is comprised of eight RECs that are responsible for regional development and integration and assist with African Peace and Security Architecture initiatives. At a national level, member states are responsible for their own policies and constitutions overseeing their political, economic and administrative governance.[27]

This section identifies and addresses the crucial frameworks and initiatives applicable to governance on a continental level.

1.3.1   The African Union

The main political body across the continent is the AU, of which all 55 African states are members. It was officially launched in 2002 after its predecessor [the Organisation of African Unity (OAU)] decided to create a new continental organisation that focusses on "increased cooperation and integration of African states to drive Africa's economic growth and economic development".[28]

The AU is currently developing three continental financial institutions to facilitate an integrated and well-regulated African economy, specifically: the African Central Bank, the African Investment Bank and the African Monetary Fund.

Legal and judicial aspects of the AU are handled by the AU Commission on Human Rights and Peoples' Rights (ACHPR), the African Court on Human and Peoples' Rights (AfCHPR), the AU Commission on International Law (AUCIL), the AU Advisory Board on Corruption (AUABC) and the ACERWC.[29]

---

[27] African Peer Review Mechanism, The African Governance Report, South Africa, 2019.

[28] The African Union Commission, *Agenda 2063: The Africa We Want*, https://au.int/en/agenda 2063/overview, accessed: 21 January 2020.

[29] Ibid.

The eight RECs are essentially regional groupings of African states,[30] with the general purpose of facilitating regional economic integration within Africa and the wider African Economic Community (AEC) (see Table 1.2).[31]

1.3.2 African Policy Framework Relevant to Governance

The basis of the governance framework within the AU is the Constitutive Act of the AU, which "defines the establishment, objectives, and principles of the AU and the major implementing organs".[32] The document reflects "global values, principals, and norms relating to human rights, sovereignty, peace and security, good neighbourliness, cultural and socio-economic values and international cooperation that are also contained in the Purposes and Principles of the UN as stated in the Charter of the United Nations Organization"[33] and forms the foundation for policies and institutional frameworks. In addition to the Constitutive Act, the AU has a set of Shared Values relating to democracy and governance, upholding the rule of law, protecting human rights, African development and integration, and promoting peace and security.[34] These Shared Values form common goals that steer the decisions and actions of member states.

The AU organs and instruments focused on the realization of the AU Shared Values relating to democracy and good governance include:

- African Charter of Democracy, Elections and Governance (ACDEG),
- OAU/AU Declaration on Principles Governing Democratic Elections,
- New Partnership for Africa's Development (NEPAD),
- Declaration on Democracy, Political, Economic and Corporate Governance and
- AU Convention on Preventing and Combating Corruption.[35]

---

[30] For an in-depth analysis of the African space sector by Regional Economic Community, see: A. Froehlich and A. Siebrits, "African Union Member States: National Space Infrastructure, Activities, and Capabilities", in *Space Supporting Africa Volume 1: A Primary Needs Approach and African's Emerging Space Middle Powers*, A. Froehlich and A. Siebrits (Cham: Springer, 2019), 191–272.

[31] The African Union Commission, *Regional Economic* Communities, https://au.int/en/organs/recs, accessed: 21 January 2020.

[32] African Peer Review Mechanism, *The African Governance Report—Promoting African Union Shared Values*, South Africa, 2019.

[33] Ibid.

[34] Ibid.

[35] Ibid.

Table 1.2 Regional Economic Communities (RECs) within the AU

Name	Acronym	Members	Primary purpose	Aims and objectives
Arab Maghreb Union	AMU	Algeria, Libya, Mauritania, Morocco, Tunisia	Improving relations between member states, encourage prosperity, create inclusive policies to promote trade, investment and free movement of people within the region, protect the national rights of its member states	Promote close diplomatic ties and dialogue between member states while safeguarding their independence; promote mechanisms for member states' industrial, commercial and social development including through common sectoral programmes; promote measures to support Islamic values and the safeguarding of the National Arabic identity through mechanisms such as cultural exchange, research and education programmes[a]
Common Market for Eastern and Southern Africa	COMESA	Burundi, Comoros, Democratic Republic of Congo, Djibouti, Egypt, Eritrea, Ethiopia, Kenya, Libya, Madagascar, Malawi, Mauritius, Rwanda, Seychelles, Sudan, Swaziland, Uganda, Zambia, Zimbabwe	Create a free trade zone within the region	Attain sustainable growth and development of member states; promote joint development in all fields of economic activity; cooperate in the creation of an enabling environment for foreign, cross-border and domestic investment; promote peace, security and stability among the member states; and strengthen relations between the Common Market and the rest of the world[a]

(continued)

Table 1.2 (continued)

Name	Acronym	Members	Primary purpose	Aims and objectives
Community of Sahel-Saharan States	CEN-SAD	Benin, Burkina Faso, Cabo Verde, Central African Republic, Chad, Comoros, Côte d'Ivoire, Djibouti, Egypt, Eritrea, Gambia, Ghana, Guinea, Guinea Bissau, Kenya, Liberia, Libya, Mali, Mauritania, Morocco, Niger, Nigeria, São Tomé and Príncipe, Senegal, Sierra Leone, Somalia, Sudan, Togo, Tunisia	Encourage and promote economic, political, cultural and social integration within its member states	Establish a comprehensive economic union with a particular focus in the agricultural, industrial, social, cultural and energy fields; adopt measures to promote free movement of individuals and capital; promote measures to encourage foreign trade, transportation and telecommunications among member states; Promote measures to coordinate educational systems; Promote cooperation in cultural, scientific and technical fields[a]
East African Community	EAC	Burundi, Kenya, Rwanda, Uganda, Tanzania	Establish a Common Market between member states (est. in 2010) as well as a Customs Union (est. in 2005) and a monetary union (in progress) to become a political federation	Develop policies and programmes aimed at widening and deepening co-operation among the partner states in political, economic, social and cultural fields, research and technology, defence, security, legal and judicial affairs, for their mutual benefit[a]
Economic Community of Central African States	ECCAS	Angola, Burundi, Cameroon, Central African Republic, Chad, Congo, Democratic Republic of Congo, Equatorial Guinea, Gabon, São Tomé and Príncipe	Promoting socio-economic development within the region and improve the living conditions of the general population	Achieve collective autonomy; raise the standard of living of its populations; maintain economic stability through harmonious cooperation[a]
Economic Community of West African States	ECOWAS	Benin, Burkina Faso, Cabo Verde, Côte d'Ivoire, Gambia, Ghana, Guinea, Guinea Bissau, Liberia, Mali, Niger, Nigeria, Senegal, Sierre Leone, Togo	Promote economic integration in all fields of economic activity	Promote cooperation and integration in the region, leading to the establishment of an economic union in West Africa in order to raise the living standards of its peoples; maintain and enhance economic stability, foster relations among member states and contribute to the progress and development of the African continent[a]

(continued)

Table 1.2 (continued)

Name	Acronym	Members	Primary purpose	Aims and objectives
Intergovernmental Authority on Development	IGAD	Djibouti, Eritrea, Ethiopia, Kenya, Somalia, South Sudan, Sudan, Uganda	Represent the interest of states in the Eastern Africa region, promote greater economic and political cooperation within the region and address peace and security issues	Promoting joint development strategies; harmonising member states' policies; achieving regional food security; initiating sustainable development of natural resources; promoting peace and stability in the sub-region; mobilising resources for the implementation of programmes within the framework[a]
Southern African Development Community	SADC	Angola, Botswana, Democratic Republic of Congo, Lesotho, Madagascar, Malawi, Mauritius, Mozambique, Namibia, Seychelles, South Africa, Eswatini, Tanzania, Zambia, Zimbabwe	Economic integration and development within the Southern Africa region	Promoting sustainable and economic growth and development; promoting common political values and systems; consolidating democracy, peace, security and stability; maximising productive employment and use of resources; achieving sustainable use of natural resources and effective protection of the environment; combating HIV/AIDS and other diseases[a]

[a]The African Union Commission, *Regional Economic Communities*, https://au.int/en/organs/recs, accessed: 21 January 2020

AU instruments aimed at upholding the rule of law, protecting and promoting human rights and encouraging inclusivity throughout the continent include:

- African Charter of Human Rights and Peoples' Rights,
- Kigali Declaration on Human Rights in Africa,
- Protocol on the Statute of the African Court of Justice and Human Rights,
- Protocol to the African Charter setting up the African Court on Human and Peoples' Rights,
- Protocol to the African Charter on Human and Peoples' Rights on the Rights of Women in Africa,
- Solemn Declaration on Gender Equality in Africa,
- African Charter on the Rights and Welfare of the Child,
- African Youth Charter and
- Draft Protocol to the African Charter on Human and Peoples' Rights on the Rights of Persons with Disabilities in Africa.[36]

Cooperation between the AU, RECs and the Coordinating Mechanisms of the Regional Standby Brigades is mandated via the Constitutive Act to ensure quick, coordinated action is available where a peace or security threat arises. Other AU instruments centred on peace and security in Africa include: the Protocol Relating to the Establishment of the Peace and Security Council of the AU, the AU Post-Conflict and Reconstruction Policy Framework, the Conference on Stability, Security, Development and Democracy (CSSDCA), the AU Convention Governing the Specific Aspects of Refugee Problems in Africa, and the AU Convention for the Protection and Assistance of Internally Displaced Persons in Africa.[37]

1.3.3 New Partnership for Africa's Development

In 2001, Algeria, Egypt, Nigeria, Senegal and South Africa established the New Partnership for Africa's Development (NEPAD) through the Millennium Africa Recovery Plan (MAP) and the Omega Plan for Africa. The NEPAD Secretariat was created and made responsible for the coordination of NEPAD initiatives and projects. In 2002, NEPAD was endorsed as an instrument of the AU at the Inaugural Summit. The NEPAD Secretariat was replaced by the NEPAD Agency in 2010 as part of integrating the programme with AU structures and processes.[38]

NEPAD is a flagship socio-economic development programme and has four core goals: (i) eradicate poverty, (ii) encourage and support sustainable growth and development, (iii) realise Africa's integration into the world economy, and (iv) accelerate the empowerment of women. NEPAD is responsible for the development of

---

[36]Ibid.

[37]Ibid.

[38]African Union Development Agency (AUDA-NEPAD), *History and Interactive Timeline*, www.nepad.org/history, accessed: 21 January 2020.

continent-wide projects and initiatives and aids in securing funding for these projects and liaising with member states, RECs and the international community to implement such projects.[39] The vision and mission of NEPAD is shown below:

- **Vision**: "Build an integrated, prosperous and peaceful Africa driven by its own citizens and representing a dynamic force in the global arena."[40]
- **Mission**: "Work with African countries, both individually and collectively towards sustainable growth and development."[41]

NEPAD focusses on four core thematic areas, which are listed below:

1. Natural Resources Governance and Food Security, which includes:

   (a) Comprehensive Africa Agriculture Development Programme (CAADP) and
   (b) Environmental Governance and Climate Change.

2. Regional Integration, Infrastructure and Trade
3. Industrialisation, Science, Technology and Innovation
4. Human Capital Development.

The numerous projects and programmes implemented through NEPAD endeavour to reduce the infrastructure gaps within Africa to allow for further development within the continent, specifically in the energy, transport, water and information and communications technology (ICT) sectors. The extensive work done under the CAADP initiative has resulted in accelerated growth throughout Africa, which, in turn, has increased food security and resilience. A special focus is placed on capacity-building and human capital development within all the programmes, which are supplemented with educational and vocational education drives. Such drives increase the knowledge base within the continent and ensure inclusivity in the economy. Since its inception, NEPAD has:

- directly impacted 1.2 million women in business, microfinance, agriculture, nutrition and ICT in 38 African countries,
- helped increase public agricultural expenditures by more than 7% on average per year through the CAADP, which 47 African countries have signed,
- constructed more than 16,000 km of road networks and 3,500 km of power transmission lines thereby increasing regional integration and providing new infrastructure,
- created a legislative framework, the AU Model Law on Medical Products Regulation, which overcomes challenges in the sector by harmonizing specifications, requirements and processes, and has been adapted and implemented in the national legislation of twelve African countries, thereby increasing the availability and access to safe medicines,

---

[39] African Union Development Agency (AUDA-NEPAD), *NEPAD in Brief*, Brochure available for download at: www.nepad.org/file-download/download/public/15083, accessed: 21 January 2020.
[40] Ibid.
[41] Ibid.

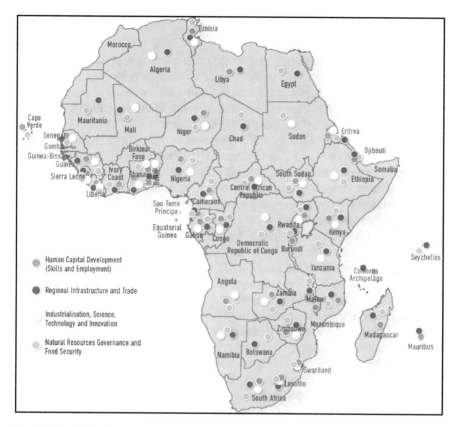

Fig. 1.4 NEPAD's footprint across Africa (NEPAD, *Annual Results based Report 2017*, South Africa, 2018)

- achieved the commitment of 85 million hectares of degraded, deforested land for restoration programmes,
- installed fibre optic Internet in 17 African countries,
- empowered over 530,000 women and youth through vocational education training programmes and business opportunities,
- created 112,900 direct jobs and 49,400 indirect jobs through the implementation and operation of multinational infrastructure projects.[42]

The NEPAD Agency assessed the Agenda 2063 FTYIP and ensured its interventions were aligned with the objectives identified within the implementation plan in order to assist Africa in its realization of Agenda 2063. The footprint of the agency currently extends to 52 of the 55 African member states, as shown in Fig. 1.4.[43]

---

[42] African Union Development Agency (AUDA-NEPAD), Results at a Glance, www.nepad.org/who-we-are/results-at-glance, accessed: 21 January 2020.

[43] NEPAD, *Annual Results based Report 2017*, South Africa, 2018.

1.3.4 African Peer Review Mechanism

The African Peer Review Mechanism (APRM) is an instrument of the AU that fosters systematic peer learning and self-monitoring of governance principles to voluntarily acceded member states. It was envisioned in the original outlining document for the NEPAD, which stated that "development is impossible in the absence of true democracy, respect for human rights, peace and good governance"[44] and was established in 2003 following the NEPAD founding document the "Declaration on Democracy, Political, Economic and Corporate Governance".[45] The vision of the APRM is "full actualization of transformative leadership" within Africa and the mission is to "promote the African Union's ideals and shared values of democratic governance and inclusive development by encouraging all member states of the Union to collaborate and voluntarily participate in the home-grown, credible, rigorous, independent and self-driven peer review process and the implementation of its recommendations".[46]

The core purpose of the APRM is "to foster the adoption of policies, standards and practices that lead to political stability, high economic growth, sustainable development and accelerated sub-regional and continental economic integration through sharing of experiences and reinforcement of successful and best practices, including identifying deficiencies and assessing the needs for capacity building".[47]

Member states perform self-assessments concerning all aspects of governance and socio-economic development, across all levels of government, private sector, civil society and the media. The APRM has identified four core thematic areas of focus: (i) democratic and political governance, (ii) economic governance and management, (iii) socio-economic governance, and (iv) corporate governance.[48]

There are four different types of reviews performed by the APRM: a "base review" performed when a new country accedes to the APRM, "periodic reviews" that are implemented on a four year cycle, "requested reviews" that are specifically requested by member states, and lastly, reviews commissioned by the APR Forum (the highest decision-making body within the APRM comprising a Committee of Participating Heads of State and Government of the participating member states) in countries where political and/or economic crises seem imminent.[49] At the time of publication, 37 countries were members of the APRM, specifically: Algeria, Angola, Benin, Burkina Faso, Cameroon, Chad, Republic of Congo, Djibouti, Egypt, Ethiopia, Equatorial Guinea, Gabon, Gambia, Ghana, Ivory Coast, Kenya, Lesotho, Liberia, Malawi, Mali, Mauritania, Mauritius, Mozambique, Namibia, Niger, Nigeria, Rwanda, São Tomé and Príncipe, Senegal, Sierra Leone, South Africa, Sudan, Tanzania, Togo,

---

[44] African Peer Review Mechanism, *Strategic Plan 2016–2020*, South Africa, 2016.

[45] African Peer Review Mechanism, *About the APRM*, www.aprm-au.org/page-about/, accessed: 21 January 2020.

[46] Ibid.

[47] Ibid.

[48] Ibid.

[49] The African Union Commission, African Peer Review Mechanism (APRM), https://au.int/en/org ans/aprm, accessed: 21 January 2020.

Tunisia, Uganda and Zambia. Out of the 37 countries included in the APRM, peer reviews have been performed on 21.[50]

1.3.4.1   APRM Structure

The structure of the APRM on a continental level consists of the APR Forum, the APR Secretariat, the APR Panel and various APR Teams. The APR Forum is mandated with appointing the APR Panel and its chairperson, communicating the recommendations from country review reports to the appropriate Head of State of a government and effecting changes in the reviewed country through peer dialogue and persuasion. The APR Forum is responsible for submitting APRM reports to the AU structures and defining the rules and code of conduct for the APRM and securing funding.[51] The APR Forum identified strategic partnerships to facilitate its functions, including: the African Capacity Building Foundation (ACBF), the African Development Bank (AfDB), the Mo Ibrahim Foundation (MIF), the UN Economic Commission for Africa (UNECA), the Office of the Special Advisor on Africa (OSAA) and the UN Development Programme (UNDP).[52]

The APR Secretariat consists of staff and consultants from various African countries and "provides technical, coordinating and administrative support services for the APRM".[53] These services include preparing panel meetings, summits and regional workshops, planning Country Review Missions and recommending specific experts for APR Teams, sharing findings, preparing background documents for APR Teams and providing assistance to participating member states. Professor Eddy Maloka is currently the chief executive officer of the APR Secretariat and works closely with National Commissions throughout the continent.[54]

The APRM process is directed and overseen by the APR Panel, which consists of between five and seven Eminent Persons with the purpose of "ensuring the independence, professionalism and credibility of the process".[55] The APR Panel is instrumental in configuring APR Teams and recommends various African experts or institutions to assist with assessments. It is also responsible for analysing country review reports and making recommendations to the APR Forum.[56]

---

[50] African Peer Review Mechanism, *Continental Presence*, www.aprm-au.org/map-areas/, accessed: 21 January 2020.

[51] African Peer Review Mechanism, *APR Forum*, www.aprm-au.org/apr-forum/, accessed: 21 January 2020.

[52] African Peer Review Mechanism, *Strategic Partners*, www.aprm-au.org/partners/, accessed: 21 January 2020.

[53] African Peer Review Mechanism, *APR Secretariat*, www.aprm-au.org/apr-secretariat/, accessed: 21 January 2020.

[54] Ibid.

[55] African Peer Review Mechanism, *APR Panel of Eminent Persons*, www.aprm-au.org/apr-panel/, accessed: 21 January 2020.

[56] Ibid.

The success of the APRM process relies on participating countries establishing relevant structures and frameworks to ensure effective implementation of Country Reviews and recommendations. In this regard, the APRM has compiled a Country Guidelines document that informs governments on the necessary actions to prepare for and participate in the APRM. The core aspects involve: establishing a national focal point and national commission, appointing a national APR secretariat, engaging with technical research institutions, and implementing an independent, well-defined budgetary framework.[57]

It is recommended that the national focal point is established at a ministerial level that reports directly to the Head of State or government and that it is well integrated and coordinated with existing policy making and planning processes. The national APRM commission is an autonomous body responsible for ensuring the credibility and independence of the APRM process and providing governments with direction specific to policy creation and implementation of the APRM. The Country Guidelines require that the national commission have "clear written terms of reference for operation" and should "offer a microcosm of the nation"[58] with regard to its composition, including both state and non-state actors to realise a diverse and representative committee (where youth, women groups, civil society, parliament, marginalized groups and the media are represented).[59]

The Country Guidelines require the establishment of a national APR secretariat responsible for providing support to the national commission in performing its responsibilities. The secretariat should be separate from government entities and sustained by an independent budget. National secretariats communicate directly between the national commissions and the continental APRM Secretariat and are required to assist with engaging and tasking local technical research institutions that are responsible for completing the APRM Questionnaire. Additionally, national APR secretariats must collate national data and convey the views of the general public using both qualitative and quantitative methods that are externally validated.[60]

1.3.4.2 Key Bottlenecks Hindering African Development

Professor Eddy Maloka championed the preparation of the 2017 report titled "The Major Bottlenecks Facing Africa", in which the APRM, along with the South African Institute for International Affairs (SAIIA) investigated how the APRM could help overcome the major barriers currently challenging Africa's development. In the foreword of the report, he states that "no amount of technology or investment will address

---

[57] African Peer Review Mechanism, *National Level*, www.aprm-au.org/national-level/, accessed: 21 January 2020.

[58] African Peer Review Mechanism, *National Level—National Commission*, www.aprm-au.org/national-level/#1497512220188-c7875e9e-fa9f, accessed: 21 January 2020.

[59] Ibid.

[60] African Peer Review Mechanism, *National Level—Technical Research Institutions*, www.aprm-au.org/national-level/#1497512994752-c2021e88-3a1a, accessed: 21 January 2020.

these challenges until and unless the entire effort is underpinned by some of the most fundamental principles of good governance".[61] The investigation for the report was established after the president of Uganda, President Museveni, identified the main bottlenecks required to be overcome for Africa to fully realise sustainable transformation. He requested that the bottlenecks be included within the APRM tool. The eleven bottlenecks were increased to thirteen following Summit deliberations and are listed below[62]:

1. Ideological disorientation,
2. Interference with the private sector,
3. Under-developed infrastructure,
4. Weak states, especially weak institutions,
5. Fragmented markets, market access and expansion,
6. Lack of industrialisation and low value addition,
7. Under-development of human resources,
8. Under-development of agriculture,
9. Under-development of the services sector,
10. A non-responsive civil service,
11. Attacks on democracy and governance,
12. Domestic resource mobilisation and
13. Structural inequalities in access to opportunities.

The research for the report involved: the 17 Country Review Reports already published by the APRM, consultation with institutional bodies, the examination of reports and academic literature, and the implementation of an online questionnaire with various stakeholders involved in African governance. The major over-reaching findings of the study concluded that "the APRM offers a sound methodology and framework to achieve its goals, its full actualization requires a number of small but collectively significant reforms".[63] The subsequent steps identified by the study involve the development and trial of indicators within the APRM framework. The findings may then be integrated into the country self-assessment questionnaires and review processes to better monitor the progress made in addressing the 13 bottlenecks.

1.3.5 African Continental Free Trade Area

On 21 March 2018, the African Continental Free Trade Area (AfCFTA) Agreement was signed at the AU Assembly in Rwanda by 54 of the African states (only Eritrea has not signed the agreement). The Agreement establishing the AfCFTA entered into force on 30 May 2019. The AfCFTA intends to increase intra-African trade by

---

[61] African Peer Review Mechanism Secretariat, *Major Bottlenecks Facing Africa*, South Africa, October 2017.

[62] Ibid.

[63] Ibid., p. 8.

expanding the free trade areas already established in the eight RECs to create a single continental market for goods and services.[64] In the exact words of the Agreement, the first objective is to "create a single market for goods, services, facilitated by movement of persons in order to deepen the economic integration of the African continent and in accordance with the Pan African Vision of 'An integrated, prosperous and peaceful Africa' enshrined in Agenda 2063".[65]

Promoting and increasing intra-African trade will provide an economic boost to African countries. Between 2017 and 2018, Africa's global exports increased by 22% but intra-African exports increased by just 1%. Intra-African exports in 2018 were valued at US$74 billion, however this value only constitutes 15% of Africa's world exports. Improving the distribution of trade between African countries is an additional benefit of the AfCFTA. Currently intra-African trade is dominated by South Africa, which accounts for 34% of intra-Africa exports and 20% of intra-Africa imports.[66]

Currently 28 African countries have ratified and deposited the Agreement[67] that comprises protocols on trade in goods, trade in services and dispute settlement. The next phase of AfCFTA negotiations will define additional protocols concerned with intellectual property rights, investments, and competition policies. The trade in goods protocol facilitates the transit and trade of goods within the continent and eliminates duties and prohibitive restrictions on imports. It also ensures that intra-African imports shall be treated no less favourably than domestic goods, removes non-tariff barriers and supports product standardization and regulation.[68] Removing import duties and lowering tariffs is expected to increase intra-African trade substantially. The UNECA estimates that removing import duties will increase trade by 52.3% and that reducing non-tariff barriers may increase trade by as much as 100%. Before the AfCFTA, average tariffs were 6.1% meaning that, for many businesses, exporting outside of Africa was cheaper than exporting within the continent.[69]

The protocol on trade in services regulates intra-African services by providing standardization and certification of service suppliers. Similar to the protocol on trade

---

[64]tralac Trade Law Centre, *The African continental free trade area: A tralac guide*, tralac NPC, November 2019, www.tralac.org/documents/resources/booklets/3028-afcfta-a-tralac-guide-6th-edition-november-2019/file.html, accessed: 27 April 2020.

[65]African Union, *Agreement establishing the African Continental Free Trade Area*, 10th Extraordinary Summit of the AU Assembly, March 2018, Rwanda.

[66]tralac Trade Law Centre, *The African continental free trade area: A tralac guide*, tralac NPC, November 2019.

[67]African Union, *List of countries which have signed, ratified/acceded to the agreement establishing the African continental free trade area*, African Union, Ethiopia, 8 October 2019.

[68]African Trade Policy Centre, *African Continental Free Trade Area: Questions & Answers*, African Union Commission, March 2018, https://au.int/sites/default/files/documents/36085-doc-qa_cfta_en_rev15march.pdf, accessed: 27 April 2020.

[69]United Nations Economic Commission for Africa, *African Continental Free Trade Area—Questions & Answers*, United Nations Economic Commission for Africa, www.uneca.org/publications/african-continental-free-trade-area-questions-answers, accessed: 27 April 2020.

in goods, the protocol on trade in services specifies that intra-African service suppliers shall be treated no less favourably than domestic suppliers. Lastly, the protocol aspires to liberalize and develop intra-African services sectors, while being cognizant of exceptions relating to national security.[70]

Once ratified by all 55 African states, in terms of numbers of countries the AfCFTA will be the world's largest free trade area since the World Trade Organization and will cover a combined GDP of US\$ 2.5 trillion.[71]

1.4 Challenges Facing Governance Within Africa

Volume 1 of Space Supporting Africa offers a detailed analysis of the African context and background, however, some of the facts bear repeating as they have a direct impact on the implementation of potential e-governance systems (and space-based applications as a whole).[72]

Africa is the second largest continent and accounts for approximately 20% of the Earth's land area comprising 30.3 million km^2 and consisting of 55 sovereign states.[73] Africa is unique in that across many metrics, it exhibits huge variances between its countries. For example, while the average urbanisation rate is 40% across the continent, Burundi, the least urbanised country, is only 12.2% urban, whereas Gabon, the most urban country in Africa, is 87.2% urban.[74]

Similarly, population densities across Africa also exhibit large variances. The most densely populated country, Mauritius, has a density of 627 people per square kilometre, while nine countries have densities of less than ten people per square kilometre.[75]

According to UN DESA, the population of Africa was 1.3 billion people in 2019 and is projected to reach almost 2.5 billion people by the year 2050. Africa has the youngest population in the world, with 60% under the age of 25 years in 2019.[76]

---

[70]African Trade Policy Centre, *African Continental Free Trade Area: Questions & Answers*, African Union Commission, March 2018, https://au.int/sites/default/files/documents/36085-doc-qa_cfta_en_rev15march.pdf, accessed: 27 April 2020.

[71]United Nations Economic Commission for Africa, *African Continental Free Trade Area—Questions & Answers*, United Nations Economic Commission for Africa, www.uneca.org/publications/african-continental-free-trade-area-questions-answers, accessed: 27 April 2020.

[72]See: A. Froehlich and A. Siebrits, *Space Supporting Africa Volume 1: A Primary Needs Approach and Africa's Emerging Space Middle Powers* (Cham: Springer, 2019). Also see: A. Siebrits, B. Martens, and C. Eriksen, "Initiatives for Embedding Space Applications in African Societies", in *Integrated Space for African Society: Legal and Policy Implementation of Space in African Countries*, ed. A. Froehlich (Cham: Springer, 2019), 357–373.

[73]National Geographic, "Africa: Physical Geography", 2018, www.nationalgeographic.org/encyclopedia/africa-physical-geography/, accessed: 23 November 2019.

[74]United Nations Economic Commission for Africa, "The Demographic Profile of African Countries", 2016, Addis Ababa.

[75]Ibid., p. 8.

[76]United Nations, Department of Economic and Social Affairs, Population Division, "World Population Prospects 2019: Data Booklet", ST/ESA/SER.A/424, 2019.

Fundamental challenges prohibiting good governance in African countries include: corruption, weak democratic institutions, limited data coverage and openness and a lack of civil registration and vital statistics (CRVS). These are discussed in detail in the coming sections as they are currently viewed as some of the major hurdles inhibiting good governance in Africa. Other relevant challenges include:

- a lack of technical capacity,
- sensitisation and awareness issues,
- lack of core infrastructure [including energy, water, sanitation and ICT],
- lack of funding and capital investment,
- low levels of connectivity,
- immigration and refugee crises,[77]
- repression of dissent,
- dispersed populations and varying education levels and
- diverse linguistic, cultural and political arenas.

1.4.1 Corruption

Corruption and a lack of accountability in the public sector continue to negatively impact many African countries. Corruption undermines democratic processes and weakens governance. Transparency International is an international civil organisation focused on reducing corruption on a global scale with a vision for "a world in which government, business, civil society and the daily lives of people are free of corruption".[78] The organisation publishes an annual Corruption Perception Index (CPI), which "measures the perceived levels of public sector corruption in 180 countries".[79] The CPI rates the perceived levels of corruption on a scale from zero (very corrupt) to 100 (little to no corruption). Concerningly, the 2018 CPI report reveals that more than two-thirds of countries failed to achieve a score higher than 50, with a global average of 43. The highest scoring country in 2018 was Denmark, with a score of 88, followed closely by New Zealand and Finland, which achieved 87 and 85 respectively.[80]

Corruption levels increase considerably when looking at the African continent, as highlighted in Fig. 1.5. The lowest scoring region globally is Sub-Saharan Africa (SSA), which has an average score of 32 out of 100. Furthermore, of the lowest three scoring countries in the 2018 report, two are African: South Sudan achieved a score of 13 and Somalia achieved just 10—the lowest score out of all the countries included

---

[77] See for additional discussion: N. Ringas, "MENASat-Proposal for a Space-Based Refugee Assistance Programme", in *Space Fostering African Societies: Developing the African Continent through Space, Part 1*, ed. A. Froehlich (Cham: Springer, 2020).

[78] Transparency International, *What is Transparency International*, www.transparency.org/about, accessed: 21 January 2020.

[79] Transparency International Secretariat, *2018 Corruption Perception Index*, Germany, 2019, ISBN: 978-3-96076-116-7.

[80] Ibid.

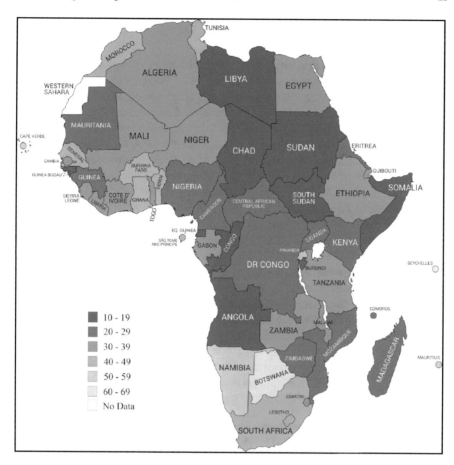

Fig. 1.5 African scores in Transparency International's 2018 Corruption Perception Index (low values indicated increased levels of corruption) (Figure recreated from the Transparency International's Corruption Perception Index 2018)

in the study.[81] The 2018 CPI report highlights a strong link between high levels of corruption and weak democracies. No country with full democratic characteristics achieved a score less than 50. This observation is re-iterated by Delia Ferreira Rubio, the chair of Transparency International, who stated that their "research makes a clear link between having a healthy democracy and successfully fighting public sector corruption".[82]

---

[81]Ibid.

[82]Ibid., p. 3.

1.4.2 Weak Democratic Institutions

Numerous countries within Africa still exhibit authoritarian governments and threats are prevalent to existing democracies as a whole. Recent examples include the Boko Haram insurgency in Nigeria, which displaced almost three million people and resulted in the death of more than fifty thousand Nigerians.[83] Similarly, the on-going Somali Civil War, which has persisted for more than three decades, has a total death toll of more than half a million people and has displaced more than two million Somalis.[84] There has been ongoing civil strife in Libya since the start of the Arab Spring movement in February 2011.[85] In the first year of the civil war, more than 435,000 people were displaced, 19,700 injured and 21,490 people were killed.[86] In 2019, the Battle of Tripoli (Libya's capital) resulted in the death of 3,000 civilians and over 5,600 being wounded.[87] The Democratic Republic of Congo endured two civil wars during the 1990s and early 2000s that killed approximately six million people.[88] In December 2018, Félix Tshisekedi won the national elections and replaced Joseph Kabila who was president for 18 years.[89] However, the Human Rights Watch notes that the "elections were marred by widespread irregularities, voter suppression, and violence".[90] Furthermore, they state that over 13 million people are in need of human-itarian assistance within the country and identified over 140 active armed militia groups in the North Kivu and South Kivu provinces.[91]

Moreover, many authoritarian leaders continue to enforce their power over numerous terms (18 African presidents have been in office for more than a decade).[92] President Teodoro Obiang Nguema Mbasogo has been the president of Equatorial

---

[83] V. Turner et al., *Nigerian refugees struggle in aftermath of Boko Haram attacks*, The UN Refugee Agency, 2019, https://www.unhcr.org/news/stories/2019/2/5c6139e74/nigerian-refugees-struggle-aftermath-boko-haram-attacks.html, accessed: 28 April 2020.

[84] UNHCR, *Somalia Situation 2017*, The UN Refugee Agency, Geneva, 2017.

[85] Frederic Wehrey, *"Our hearts are dead." After 9 years of civil war, Libyans are tired of being pawns in a geopolitical game of chess*, Time USA, Libya, February 2020, https://time.com/5779348/war-libya-global-conflict/, accessed: 20 May 2020.

[86] Mohamed A. Daw, *Libyan armed conflict 2011: Mortality, injury and population displacement*, Tripoli, Libya, 2015, https://doi.org/10.1016/j.afjem.2015.02.002.

[87] Al Jazeera Media Network, *WHO: More than 1000 killed in battle for Libya's Tripoli*, Al Jazeera Media Network, July 2019, www.aljazeera.com/news/2019/07/1000-killed-battle-libya-tripoli-190708191029535.html, accessed: 20 May 2020.

[88] BBC News, *DR Congo country profile*, BBC, January 2019, www.bbc.com/news/world-africa-13283212, accessed: 20 May 2020.

[89] The World Bank, *The World Bank in DRC—Overview*, International Bank for Reconstruction and Development, The World Bank Group, May 2020, www.worldbank.org/en/country/drc/overview, accessed: 20 May 2020.

[90] Human Rights Watch, *Democratic Republic of Congo Events in 2018*, Human Rights Watch, 2020, New York, USA, www.hrw.org/world-report/2019/country-chapters/democratic-republic-congo, accessed: 20 May 2020.

[91] Ibid.

[92] The Economist Intelligence Unit Limited, *Democracy Index 2018: Me Too?*, London, 2019.

Guinea for more than forty years since he gained power in 1979.[93] Similarly, the president of Chad, President Idriss Deby, who gained power through a violent coup, has been in office since 1990.[94] This poses a direct barrier to the freedom of African citizens and the promotion of good governance. However, 2018 saw countries such as Angola and Ethiopia demonstrate positive shifts towards better governance through the strengthening of their democratic characteristics.[95]

The Economists Intelligence Unit's (EIU) annual *Democracy Index* offers insight into the state of democracy for 165 independent states and two territories, representing almost the full global population. The EIU was created in 1946 and forms the research and analysis division of The Economist Group. The *Democracy Index* classifies countries as "full democracy", "flawed democracy", "hybrid regime", and "authoritarian regime", depending on the scores it achieves over a range of indicators across five major categories, namely: (i) electoral process and pluralism, (ii) civil liberties, (iii) functioning of government, (iv) political participation and (v) political culture.

The five interrelated categories offer a comprehensive conceptual representation of how substantive a country's democracy really is and differ slightly to other international indices or metrics that focus more on political freedom. In this regard, the EIU states that "freedom is an essential component of democracy, but not, in itself, sufficient".[96] The 2019 *Democracy Index* states that of the 167 states and territories analysed globally, only 20 achieve "full democracy" status, with a total of 53 "authoritarian regimes". Yet, 2018 witnessed an increase in political participation in almost all global regions. This signifies that globally, an increasing number of citizens are becoming actively engaged in the democratic process.

The 2018 results show that of the 44 countries in SSA, Mauritius was the only "full democracy", with seven "flawed democracies", 14 "hybrid regimes" and 22 "authoritarian regimes". Similarly, in the Middle East and North Africa region, out of 20 countries, there were no "full democracies", two "flawed democracies", four "hybrid regimes" and 14 "authoritarian regimes".[97]

While the average regional score in the *Democracy Index* increased in SSA, seven of the 15 lowest-ranked countries in the world are located in the region and nine countries exhibited a decline in their overall score. Furthermore, the regional average for the electoral process category has weakened slightly since 2017. This was heavily influenced by the Cameroonian election in October 2017, which had low voter turnout, poor security and was rife with irregularities and saw the Cameroonian president secure his seventh term in office. The SSA regional averages remain below the global *Democracy Index* averages for all categories but most noticeably in the

---

[93] BBC News, *Equatorial Guinea profile—leaders*, BBC, 2014, www.bbc.com/news/world-africa-13317176, accessed: 28 April 2020.

[94] BBC News, *Chad profile*, BBC, 2016, www.bbc.com/news/world-africa-13164688, accessed: 28 April 2020.

[95] The Economist Intelligence Unit Limited, *Democracy Index 2018: Me Too?*, London, 2019.

[96] Ibid.

[97] Ibid.

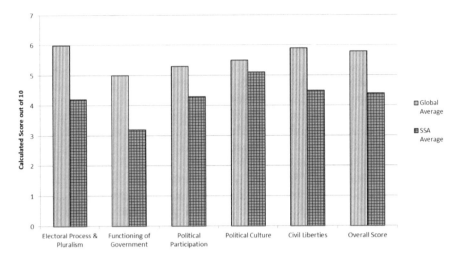

Fig. 1.6 Democracy Index average scores by category in the Sub-Saharan African region (Index score out of 10, with 10 being best) (Ibid., p. 29)

electoral process, functioning of government and civil liberties categories, as shown in Fig. 1.6.

The SSA region has consistently achieved poor scores in the civil liberties category as a result of oppression of opposition parties and stifled media content. In 2018, Tanzania placed prohibitive regulations on online content and bans were placed on opposition protests in Togo. Despite this, positive civil liberty progress was witnessed in Gambia where media freedom was promoted by the government and in Ethiopia where previously incarcerated political prisoners were released in 2018.[98]

As previously stated, Mauritius was the only country in Africa to achieve "full democracy" status. In total, eight countries in Africa achieved "flawed democracy" status, namely Botswana, Cape Verde, Ghana, Lesotho, Namibia, Senegal, South Africa and Tunisia.[99]

Freedom House is an independent international organisation founded in 1941 and is "dedicated to the expansion of freedom and democracy around the world". It aims to increase "political rights and civil liberties through a combination of analysis, advocacy, and action".[100] Freedom House creates annual *Freedom in the World* reports that evaluate political rights and civil liberties in countries around the world, based on "the assumption that freedom for all people is best achieved in liberal democratic societies".

The methodology used by Freedom House is largely based on the UN Universal Declaration of Human Rights. Data gathered from news events, academic

---

[98] The Economist Intelligence Unit Limited, *Democracy Index 2018: Me Too?*, London, 2019.

[99] Ibid.

[100] Freedom House, *Freedom in the World 2018—Methodology*, https://freedomhouse.org/report/methodology-freedom-world-2018, accessed: 3 February 2020.

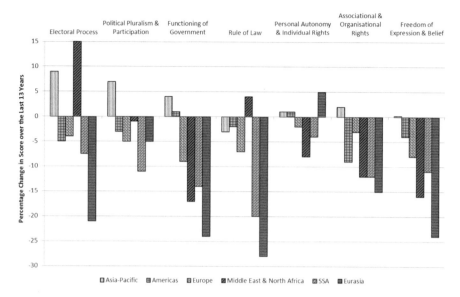

Fig. 1.7 Graph highlighting the 13 years of decline in all categories of the *Freedom in the World* Annual Report (Freedom House, *Freedom in the World* 2019, New York, 2019, p. 9, https://freedomhouse.org/sites/default/files/Feb2019_FH_FITW_2019_Report_ForWeb-compressed.pdf, accessed: 3 February 2020)

papers, reports issued by non-governmental organisations (NGOs), individual experts/professionals and on-the-ground research. Data are subsequently analysed and interpreted by more than 100 analysts and 30 advisors to determine a rating for political rights and another for civil liberties for each country/territory.

The rating for political rights assesses the electoral process, political pluralism and participation and the functioning of the government. A country's civil liberties rating is determined by assessing its freedom of expression and belief, associational and organisational rights, and the rule of law. The ratings range from one to seven, with 1.0 being the greatest degree of freedom. The average of the two ratings is used to classify a country/territory's status as Free (1.0–2.5 rating average), Partly Free (3.0–5.0), or Not Free (5.5–7.0).[101]

The 2019 report comprises 195 countries and 14 territories and reveals that young democracies have been negatively impacted as a result of corruption, anti-liberal populist movements and reductions in the rule of law. The report notes an increase in opposition movement bans and censorship of media outlets by authoritarian leaders and reveals a global decline in freedom for the 13th consecutive year, most notably in the Eurasia and SSA regions, as shown in Fig. 1.7.

The 2019 report noted that 37% of the world's population is living in Not Free countries, 24% in Partly Free and 39% in Free countries. Trends in the Middle East and North Africa region highlighted increased suppression of dissent in 2019. It was

---

[101] Ibid.

noted that political repression increased in Egypt following its last election in 2018 where potential opposition leaders were imprisoned by security forces and 97% of the vote was cast in favour of President Abdel Fattah al-Sisi. In Tunisia (which was the only Arab Spring country to avoid civil war after the 2011 uprisings and which received international praise for establishing a democratic constitution and holding free elections following the uprisings), failure to establish a Constitutional Court undermined the rule of law and legislative changes restricted freedom of association and assembly.[102] Only 4% of the Middle East and North African population are classified as Free, with 13% classified as Partly Free, and a staggering 83% as Not Free.[103]

In the SSA region, certain countries achieved significant democratic progress but threats to freedom across other countries in the region increased. Angola and Ethiopia elected new leaders who are committed to enacting important reforms. Similarly, Gambia experienced a notable democratic opening in its 2018 legislative elections, which saw numerous independent candidates from seven different parties obtaining seats.[104]

These democratic openings are shadowed by other actors in the region where political activities indicate a regression in terms of democracy. The Tanzanian government suppressed anti-government protests, promoted legislation to further its power in domestic politics and arrested prominent opposition leaders. Activities pointing to a similar shift towards a close in political activities were also observed in Uganda and Senegal. In Uganda, legislation was passed to implement taxes on social media use as well as establish new state surveillance systems. In Senegal, new regulatory barriers could limit the number of opposition parties in the upcoming elections.[105] Of the 49 African countries in the region, Freedom House rated only nine as Free. 21 countries were deemed Partly Free and 19 are Not Free—equating to 39% of the region's 1.1 billion population. Uganda moved from Partly Free to Not Free status following these aforementioned attempts to suppress freedom of expression. Conversely, Zimbabwe shifted from Not Free to Partly Free after the 2018 election that saw President Mnangagwa inaugurated.[106]

Figure 1.8 shows the status of freedom within African countries. Notably, only ten African countries are classified as free, namely: Benin, Botswana, Cape Verde, Ghana, Mauritius, Namibia, São Tomé and Príncipe, Senegal, South Africa and Tunisia.

---

[102] Freedom House, *Freedom in the World* 2019.

[103] Ibid.

[104] Ibid.

[105] Ibid.

[106] Ibid.

Fig. 1.8 Freedom classification of African countries as per the *Freedom in the World* 2019 Report (Figure created from data contained within the Freedom House *Freedom in the World* 2019 Report)

1.4.3 Limited Data Coverage and Openness

While progress has been made in the establishment of frameworks and strategies to address statistical development in Africa, statistical capacity and meaningful, contemporary data availability within the continent remains a large challenge.[107] The MIF[108] defines statistical capacity as:

> a nation's ability to collect, analyse, and disseminate high-quality data about its population and economy. Quality statistics are essential for all stages of evidence-based decision making,

---

[107] Mo Ibrahim Foundation, *African Governance* Report, 2019, Sect. 3, p. 67.

[108] Mo Ibrahim Foundation, website: https://mo.ibrahim.foundation, accessed 13 January 2020.

which include: monitoring social and economic indicators, allocating political representation and government resources, guiding private sector investment and information for the international donor community for programme design and policy formulation.[109]

Thus, statistical capacity is a critical prerequisite for good governance. This is reinforced by J. Kahimbaara in his *Report on the Pan-African Statistical Institute* wherein he states that official statistics have two major roles, "the first is to inform and monitor development. The second is to facilitate good governance."[110] He continues by stressing the importance of statistical literacy within the population to ensure that individuals understand policies, strategies, targets and other decisions that influence the socio-economic environment of a country. This statistical literacy promotes informed debate and increases participation in government transparency and accountability.[111]

Recent noticeable developments focusing on statistical development throughout Africa include the African Charter on Statistics (2009), the Praia Group on Governance Statistics (2015), the Africa Data Consensus (2015) and the Pan-African Statistics programme.[112] The African Charter on Statistics defines a set of professional ethics applicable to all statistics in the African environment and focusses on six core principles, namely:

1. Scientific independence of statistics and data,
2. Quality of statistics,
3. Mandate for data collection and resources,
4. Dissemination of statistics,
5. Protection of individual data, information sources and respondents and
6. Coordination and cooperation.

The charter has been signed by 33 African states but ratified by just 23.[113] The Praia Group was created to define well-documented data collection methodologies and techniques focused around governance statistics. These will be compiled into a handbook that will assist the National Statistical Offices (NSOs) of member states.[114]

The African Data Consensus was created in March 2015 at the 8th AU-ECA Joint Conference of Ministers, after African Heads of State requested UNECA, AUC, AfDB and UNDP to "discuss the data revolution in Africa and its implications for

---

[109]Mo Ibrahim Foundation, *African Governance* Report, 2019, Sect. 3, p. 69.

[110]J. Kahimbaara, *Report on the Pan-African Statistical Institute*, Programme implemented by Expertise France for the African-EU Partnership, Paris, France, https://au.int/en/ea/statistics/statafric, accessed: 27 November 2019.

[111]Ibid.

[112]Mo Ibrahim Foundation, *African Governance* Report, 2019, Sect. 3, p. 67.

[113]Ibid.

[114]Ibid.

AU's Agenda 2063 and the post-2015 development agenda"[115] at the 23rd Ordinary Session of the AU.[116]

The document provides a roadmap for realizing the African data revolution by including state and non-state actors in the process of data production and dissemination, and stresses the importance of population and housing censuses (PHCs) as well as frequent labour force surveys. The road map identifies six critical actions to help advance the data revolution:

1. Securing political commitment,
2. Building the evidence base,
3. Embedding the data revolution in African countries,
4. Ensuring financing and sustainability,
5. Building capacity and skills and
6. Building partnerships and synergies at national, continental and global levels.[117]

The Pan-African Statistics programme forms part of the larger Pan-African Programme implemented via the Africa-EU partnership and intends to "improve the production and dissemination of quality statistics in Africa"[118] by building on the Strategy for the Harmonisation of Statistics in Africa (SHaSA) and the aforementioned African Charter on Statistics, as well as to provide support to the new AU Institute for Statistics in Africa (STATAFRIC).[119] The programme has a budget of €6.8 million and has the following objectives:

- "[Increasing the] availability of statistical information for decision-making and policy monitoring of African integration,
- [Implementing] better frameworks for collecting, producing and disseminating harmonised statistics in Africa, and
- Institutional capacity building for good quality official statistics that underpin the African integration process and measure progress towards global goals."[120]

Substantial challenges that hinder meaningful data collection and statistics production, as highlighted by the MIF, include: lack of governmental statistical capacity, inadequate funding of NSOs, lack of independence of NSOs, limited data availability and limited data accessibility due to late publication and varying reporting formats.[121] Furthermore, poor data coverage coupled with limited statistical data openness compounds the problem. This is highlighted in Table 1.3, which shows the

---

[115] AU-ECA, *African Data Consensus*, Eighth Joint Annual Meetings of the African Union, Ethiopia, 2015, www.uneca.org/sites/default/files/PageAttachments/final_adc_-_english.pdf, accessed: 3 February 2020.

[116] Ibid.

[117] Ibid.

[118] European Commission and the Pan-African Programme, *Pan African Statistics*, Brochure, www.africa-eu-partnership.org/sites/default/files/pan_african_statistics_factsheet.pdf, accessed: 3 February 2020.

[119] Ibid.

[120] Ibid.

[121] Mo Ibrahim Foundation, *African Governance Report*, 2019.

Table 1.3 African Data Coverage and Openness Averages over 2015–2018 as per Open Data Watch's Open Data Inventory (ODIN) ODIN expressed as percentages

Data categories	Data coverage (%)					Data openness (%)				
	2015	2016	2017	2018	Trend	2015	2016	2017	2018	Trend
Social statistics sub-score	**36.1**	**30.8**	**25.6**	**29.2**	**−7.0**	**20.7**	**34.8**	**31.6**	**35.6**	**14.9**
Education outcomes	40.8	33.8	25.8	26.7	−14.1	21.3	38.3	33.4	35.7	14.4
Education facilities	35.1	31.0	19.6	22.4	−12.7	19.6	31.3	23.0	26.5	7.0
Population and vital statistics	47.4	36.9	32.6	37.5	−9.9	25.7	35.5	35.8	38.8	13.0
Health outcomes	35.3	32.6	25.8	28.9	−6.3	20.6	36.4	34.0	40.4	19.8
Gender statistics	28.7	23.4	21.1	24.5	−4.1	17.0	31.5	34.6	40.6	23.6
Poverty and income	29.8	21.1	21.0	28.1	−1.7	16.6	35.3	37.6	40.0	23.4
Reproductive health	35.9	32.1	30.7	35.0	−0.9	24.5	41.7	37.4	45.1	20.6
Crime and justice	−	−	12.7	15.9	3.2	−	−	15.8	17.3	1.5
Health facilities	36.0	35.3	41.5	43.6	7.6	19.8	27.9	33.0	35.5	15.7
Economic and Financial Statistics Sub-Score	**49.7**	**44.8**	**42.1**	**47.9**	**−1.8**	**20.6**	**34.1**	**34.3**	**39.7**	**19.1**
International trade	79.1	55.0	57.7	57.5	−21.6	21.5	37.7	36.4	39.8	18.3
National accounts	49.5	38.3	38.5	40.1	−9.4	23.6	39.1	40.6	44.1	20.5
Labour	41.0	33.7	29.1	34.2	−6.8	23.2	38.5	36.0	41.6	18.4
Money and banking	55.7	51.8	41.7	51.0	−4.7	17.4	27.0	26.6	35.5	18.1
Price indexes	40.5	34.8	34.1	47.7	7.2	19.8	37.4	43.4	48.2	28.4
Government finance	40.4	53.2	50.0	49.7	9.3	18.3	29.8	32.0	34.1	15.8
Balance of payments	56.4	60.3	56.0	66.0	9.6	20.2	28.9	25.0	34.7	14.5
Environment statistics sub-score	**22.5**	**19.4**	**19.3**	**24.0**	**1.5**	**16.1**	**22.0**	**22.8**	**26.8**	**10.6**
Energy use	14.5	9.1	4.4	7.7	−6.8	13.8	10.4	6.4	10.2	−3.6
Land use	25.5	21.9	23.6	23.9	−1.7	16.0	25.7	30.8	33.3	17.3
Built environment	37.6	37.4	34.7	37.0	−0.6	22.8	38.1	39.4	43.9	21.1
Pollution	6.4	4.5	5.5	8.4	2.0	8.3	8.7	9.6	13.7	5.4
Resource use	28.3	24.0	28.6	39.8	11.5	19.8	26.8	28.0	32.9	13.1
Total across all categories	**35.0**	**30.6**	**28.0**	**32.7**	**−2.2**	**19.1**	**30.3**	**29.6**	**34.0**	**14.9**

Data from Open Data Watch's ODIN portal and summarised in the Mo Ibrahim Foundation *African Governance Report*, 2019, pp. 74, 75.

2015–2018 African averages for data coverage and openness as a percentage across 21 different data categories (grouped into three different clusters), as per Open Data Watch's Open Data Inventory (ODIN).[122]

The MIF's Ibrahim Index of African Governance (IIAG) 2018 African Governance Report (published in 2019 and discussed in detail in Sect. 1.5.2) noted that the 2017 African average score for the *Governmental Statistical Capacity* sub-indicator was 54.5 out of a maximum score of 100, and that it has been increasing over the past decade at an annual average rate of 0.43 points. It also noted that Somalia and Libya received scores of less than 10.0 (as highlighted in Fig. 1.9). Of particular concern, 24 African countries indicated a negative trend in this sub-indicator over the 2014–2017 period indicating that the capacity of NSOs is on the decline in almost half the countries in the continent.[123]

The IIAG 2018 report found that a major challenge relating to impartial statistical data in Africa is inadequate funding, coupled with limited autonomy. The sub-indicator *Independence of National Statistical Offices* of the IIAG quantifies to what extent an NSO has "autonomy to collect data of its choosing, autonomy to publish data without prior clearance or approval from any branch of the government, and sufficient funding to collect and publish of its choosing".[124] The African average score for this sub-indicator was 33.8 out of 100 for the 2017 year, with 15 countries in Africa scoring zero (see Fig. 1.10). As such, the funding of NSOs and ensuring their independence is a fundamental area where African states need to improve in order to better facilitate governance across the continent.

1.4.4 Poor Civil Registration and Vital Statistics

An additional issue compounding the lack of data coverage is limited data availability in the form of Population and Housing Censuses (PHCs), which play an integral role in evaluating policies and strategies relating to inclusive socio-economic development, especially in developing countries with limited civil registration systems (such as for birth certificates, identity cards and death certificates). This problem is highlighted by the UN DESA Statistics Division, which states that "acquiring knowledge of the size and characteristics of a country's population on a timely basis is a prerequisite to socio-economic planning and informed decision making".[125] A common issue in the African context is infrequent PHC programmes. In the period 2009–2018, only

---

[122]Open Data Watch is a non-profit organisation of data experts founded in 2013 who aim to support national statistical offices through policy advice, data support and monitoring work. The international organisation operates the Open Data Inventory (ODIN) which "assesses the coverage and openness of official statistics to help identify gaps, promote open data policies, improve access, and encourage dialogue between national statistical offices and data users. The data for 178 countries over the 2018/2019 period are available through their website: https://odin.opendatawatch.com/, accessed: 3 February 2020.

[123]Mo Ibrahim Foundation, *African Governance* Report, 2019.

[124]Ibid., p. 71.

[125]UN DESA Statistics Department, *Principles and Recommendations for a Vital Statistics System*, Statistical Papers, Series M, No. 19, Rev 3, New York, 2014.

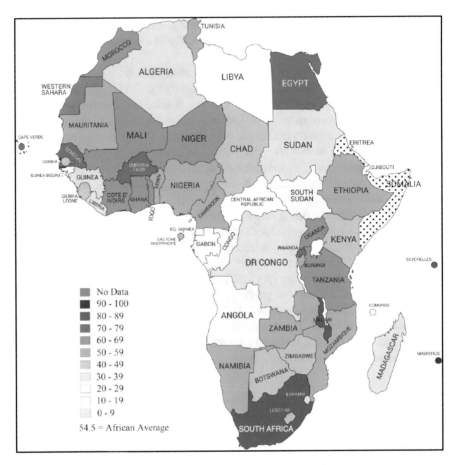

Fig. 1.9 MIF IIAG African scores for Governmental Statistical Capacity (Figure created from data contained within the MIF African Governance Report 2018)

39 African countries conducted a PHC. Although this sounds significant, in reality it means that only 54% of the continent's population lives in a country that conducted a PHC during the ten year period. The irregular nature of PHCs in Africa results in large data gaps that make measuring governance levels across African states a difficult task. For example, in 2017 data regarding the tracking of SDG progress was only available for 37.8% of the SDG indicators for African countries.[126]

Other barriers relating to meaningful data and statistics in Africa arise from differences in methodologies, data acquisition techniques and reporting formats, as well as inadequate financial support and poor data infrastructure. Moreover, late publication

---

[126]Mo Ibrahim Foundation, *African Governance* Report, 2019.

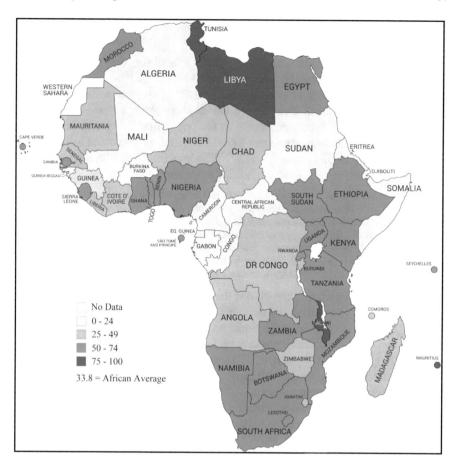

Fig. 1.10 MIF IIAG African scores for Independence of National Statistics Offices (2017) (Figure created from data contained within the MIF African Governance Report 2018. The *Independence of National Statistics Offices* sub-indicator is evaluated using Global Integrity's African Integrity Indicators available at www.globalintegrity.org/resource/aii6-xls/)

of results along with a lack of awareness and poor integration of user needs makes for limited data accessibility.[127]

As discussed above, ensuring good governance and monitoring the efficacy of governance programmes and policies depends on the availability of reliable and applicable data and statistics. One of the most important types of statistics used to evaluate governance is referred to as vital statistics, which are defined by the UN DESA Statistics Division as:

---

[127]Ibid.

"the collection of statistics on vital events in a lifetime of a person as well as relevant characteristics of the events themselves and of the person or persons concerned".[128]

Data are collected for the following events: live births, deaths, foetal deaths, marriages, divorces, annulments, separations and adoptions.[129]

One of the main sources of vital statistics is civil registration, which is defined as "the continuous, permanent, compulsory, universal recording of the occurrence and characteristics of vital events pertaining to the population".[130] Other data sources include health-care records, data from sample registration areas, and investigations into fertility and mortality through PHCs.[131] The importance of CRVS is highlighted by the UN DESA Statistics Division, which states that:

"Vital statistics and their subsequent analysis and interpretation are essential for setting targets and evaluating social and economic plans, including the monitoring of health and population intervention programmes, and the measurement of important demographic indicators of levels of living or quality of life, such as expectation of life at birth and the infant mortality rate. Vital statistics are also invaluable for planning, monitoring and evaluating various programmes such as those dealing with primary health care, social security, family planning, maternal and child health, nutrition, education, public housing and so forth".[132]

As such, civil registration constitutes the first step towards good governance of any country, as it is required for an individual to: obtain an identity card and/or passport, attend school, vote in elections, open a bank account, purchase or inherit property, own land or businesses, and attend university. The central components of a successful vital statistics system are: (i) legal registration, (ii) statistical reporting, and (iii) collection, processing, and dissemination of statistics pertaining to vital events. An overview of such a system is shown in Fig. 1.11.

In the IIAG 2018 African Governance Report, seven main difficulties impeding African countries from successfully implementing CRVS systems are identified, namely[133]:

1. Civil registration in Africa is often project-driven.
2. NSOs and national civil registration authorities exhibit little to no coordination.
3. Poor policies and frameworks relating to CRVS with low levels of commitment from governments.
4. National development frameworks do not identify CRVS as a critical focus.
5. CRVS systems are often fragmented and underutilized.
6. NSOs do not have adequate funding.
7. Inadequate policies across all levels (national, regional and continental) relating to CRVS systems.

---

[128] UN DESA Statistics Department, *Principles and Recommendations for a Vital Statistics System*, Statistical Papers, Series M, No. 19, Rev. 3, New York, 2014.

[129] UN DESA Statistics Department, *Principles and Recommendations for a Vital Statistics System*, Statistical Papers, Series M, No. 19, Rev. 3, New York, 2018.

[130] Ibid., para. 279.

[131] Ibid., para. 4.

[132] Ibid.

[133] Mo Ibrahim Foundation, *African Governance* Report, 2019.

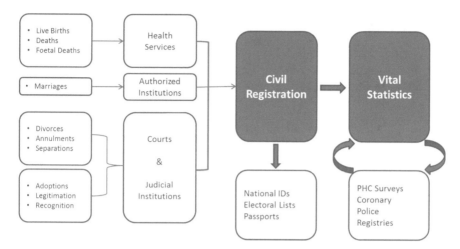

Fig. 1.11 Overview of a civil registration and vital statistics (CRVS) system (Ibid., p. 9)

According to the UN Statistics Division (UNSD) Coverage of Birth and Death Registration dataset, 42 African countries have birth registration coverage data available over the period from 2009 to 2018, with only eight countries achieving a coverage rate higher than 90%. An additional four countries indicate coverage rates higher than 90% but their data are outdated—Algeria, Libya and Tunisia have observations from 2001 and Djibouti from 2006. Only 16 countries have data sets relating to death registration coverage available and just three exhibit coverage rates greater than 90%. The worst performing country, Niger, achieved a 3.5% death coverage rate in 2018.[134]

The IIAG African Governance Report assesses the performance of African countries in this respect employing the *Civil Registration* indicator which rates a country's birth and death registration system. The African average for this indicator was 60.4 (out of 100) in 2017. However, since the implementation of the FTYIP, eight countries have seen declines in this area—most notably Malawi, which dropped by 62.5 points and the Republic of the Congo, Seychelles and Sudan who dropped 25.0 points each.[135]

Additionally, according to the 2018 ODIN dataset, African countries on average only meet 37.5% of ODIN's data coverage criteria for population and vital statistics, and 38.8% of the criteria for data openness in the same category (see Fig. 1.12 and Fig. 1.13 respectively). The African average for data coverage of population and vital statistics has decreased by 9.9% since 2015 (see Table 1.3) and eight countries

---

[134]UN Statistics Division (UNSD), *Demographic and Social Statistics*—Coverage of Birth and Death Registration dataset, available for download in Excel format at https://unstats.un.org/unsd/demographic-social/crvs/index.cshtml#coverage, accessed: 8 February 2020.

[135]Mo Ibrahim Foundation, *African Governance* Report, 2019.

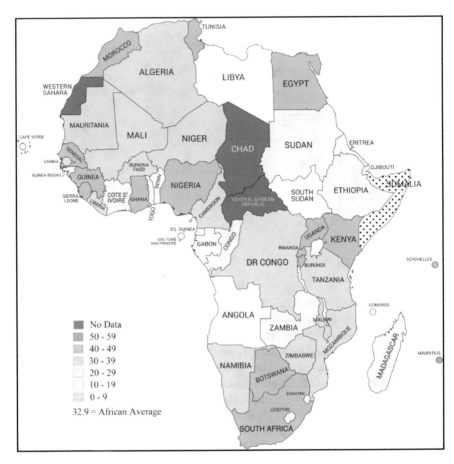

Fig. 1.12 African country scores for Data Coverage relating to ODIN's Population and Vital Statistics Data Category in 2018 (Figure developed from Open Data Watch's ODIN 2018 dataset available at https://odin.opendatawatch.com/Report/regionalProfile.)

meet none of the criteria for data coverage, namely Angola, Eswatini, Gabon, Ivory Coast, Madagascar, São Tomé and Príncipe, Somalia and Sudan.[136]

The Africa Programme on Accelerated Civil Registration and Vital Statistics (APAI-CRVS) is a regional programme committed to improving CRVS systems in Africa. APAI-CRVS was created in 2010 when the African Ministers Responsible for Civil Registration held their first conference in Addis Ababa and is now a permanent forum of the AUC.[137]

---

[136]Open Data Watch, *Open Data Inventory 2018/2019*, https://odin.opendatawatch.com/, accessed: 3 February 2020.

[137]APAI-CRVS, *Why improving civil registration and vital statistics systems in Africa is important—Making Everyone Visible in Africa!*, 2017, https://apai-crvs.org/sites/default/files/public/Making%20Everyone%20Visible_September%20EN%20_0.pdf, accessed: 3 February 2020.

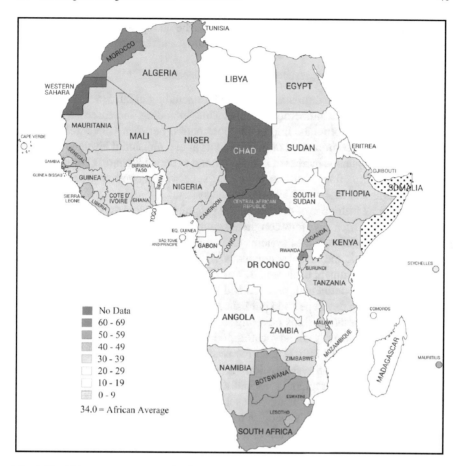

Fig. 1.13 African country scores for Data Openness relating to ODIN's Population and Vital Statistics Data in 2018 (Figure developed from Open Data Watch's ODIN 2018 dataset available at https://odin.opendatawatch.com/Report/regionalProfile)

The programme is managed by a regional APAI-CRVS Secretariat based at UNECA (and is steered by a regional CRVS Core Group comprised of partnerships with UNECA, AUC, AfDB, Secretariat of African Symposium on Statistical Development, UNICEF, World Health Organization (WHO), UN High Commissioner for Refugees (UNHCR), UN Population Fund (UNFPA), INDEPTH Network, Plan International and Partnership in Statistics for Development in the 21st Century (PARIS21)[138].

The implementation of APAI-CRVS is informed by five guiding principles:

1. "Promoting country ownership and leadership,
2. Promoting phase-based, holistic and integrated approaches,

---

[138] APAI-CRVS, *About APAI-CRVS*, www.apai-crvs.org/about-apai, accessed: 3 February 2020.

3. Promoting systematic and coordinated approaches at regional and national levels,
4. Strengthening and building capacities of national and regional CRVS institutions, and
5. Promoting innovation, research and knowledge sharing."[139]

The first phase of the implementation plan (concluding in 2015) was the preparatory phase and realized numerous achievements, including: increased capacity and education of CRVS officials, improved coordination between CRVS role players, increased political commitment of governments at the ministerial level to improving national CRVS systems, increased knowledge and experience sharing between countries, and the formation of strong partnerships between countries with developed CRVS systems and those with less developed systems.

Following its fifth year of operation, the programme noted "approximately half the countries in Africa have now completed comprehensive assessments of their civil registration and vital statistics systems and a number of them are in the process of developing strategic and costed action plans".[140] Furthermore, in 2016, the AU Heads of States and Government declared 2017 to 2026 as the "decade for repositioning of CRVS in Africa's continental, regional and national development agenda".[141]

The Core Group developed and published the first medium term plan, with costed strategic implementation steps covering 2017–2021. While the first implementation plan focused heavily on advocating, communicating and planning relating to the programme, the 2017–2021 plan focuses on intensive capacity building within countries to ensure CRVS system reform is possible. The 2017–2021 plan provides a framework that defines regional priorities and actions and establishes the associated time frames and human and financial resources required for those actions.[142]

The aim of APAI-CRVS's medium term plan is that by 2021 "all countries will have in place a comprehensive law aligned with international standards and that all African countries will have at least a 70% birth and 35% death registration coverage, along with a significant improvement in the recording of cause of death".[143]

Another important initiative is the World Bank Group's Identification for Development (ID4D), which assists countries in implementing inclusive, digitized identification systems that supplement attaining the UN SDGs. Direct impacts of a well-functioning, inclusive identification system include: access to health services, access to financial services, strengthened democracies, social and human rights protection,

---

[139] APAI-CRVS, *Why improving civil registration and vital statistics systems in Africa is important*, 2017.
[140] Ibid.
[141] Ibid.
[142] Ibid.
[143] Ibid.

and the empowerment of women and children.[144] The initiative performs assessments of the identification ecosystem within a country at its request in accordance with the *Guidelines for ID4D Diagnostics* document.

Once the assessment is completed, ID4D provides technical assistance and advisory services to governments to assist them in designing and implementing modern identification systems in accordance with international best practices. To date, ID4D has performed assessments in more than 30 countries. The technical assistance it provides includes recommendations on designs, review and development of necessary policies and legal frameworks (including data privacy legislation), integration of identification systems with appropriate governance activities, establishing evaluation programmes to monitor the implementation of the system, and ensuring and promoting end-user engagement throughout the design and implementation phases.[145]

The ID4D initiative is also focused on increasing awareness of the salience of identification systems and unifying global progress and interests in this respect. ID4D partnered with the World Bank's Global Findex in 2017, which surveyed 99 countries and revealed that almost 40% of adults in low-income countries do not have identification. ID4D is targeting the reduction of this significant number through global lobbying and sensitization. In this regard it developed the *Principles on Identification for Sustainable Development* in partnership with governments, UN agencies and private sector actors, which have been endorsed by more than 20 countries.[146]

The initiative has also produced the *Catalogue of Technical Standards for Digital Identification Systems*, which lists all relevant international standards and assists users in determining the minimum technical requirements, navigating the various standards, and identifying areas with competing standards. The ID4D initiative also facilitates peer-to-peer knowledge exchange between countries to improve global cooperation, knowledge sharing and capacity building and also manages a global dynamic research forum focused on identification systems and monitoring the global implementation thereof.[147]

Within Africa, ID4D is currently involved in the West African Regional Project, which seeks to ensure mutual recognition of identification systems in all 15 countries in the ECOWAS region (consisting of Benin, Burkina Faso, Cape Verde, Gambia, Ghana, Guinea, Guinea Bissau, Ivory Coast, Liberia, Mali, Niger, Nigeria, Senegal, Sierra Leone and Togo).

Mutual recognition across identification systems facilitates freedom of movement of citizens, increased access to services across borders, safe migration of people, and

---

[144]World Bank Group, *Identification for Development (ID4D)—Making Everyone Count*, https://pubdocs.worldbank.org/en/332831455818663406/WorldBank-Brochure-ID4D-021616.pdf, accessed: 3 February 2020.

[145]Ibid.

[146]Ibid.

[147]Ibid.

facilitates increased regional trade. To achieve this ambitious project, three core objectives must be realized, specifically[148]:

1. Develop and implement clear legal frameworks to facilitate the mutual recognition of identification systems and ensure data privacy and protection, along with well-defined technical standards for the systems and specific mandates for relevant civil institutions.
2. Assist in establishing identification systems in states where they are not present. Increase the coverage and reliability of identification systems in states which currently have their own systems in place—for example in Guinea only 60% of the population have identification and research has shown that increased coverage will support higher access rates to primary schools, vaccinations and cash transfer systems. Similarly, improving the reliability and integration of their identification system with the national social registry will assist with social programmes such as subsidized health insurance.
3. Link systems across the region to allow for self-authentication to promote freedom of movement and increase access to social transfer payments, financial transactions and crossing borders.

1.5 Current Governance Levels in Africa

This section examines the findings of the latest versions of the APRM African Governance Report and the MIF IIAG African Governance Report to determine the current governance levels in African states.

1.5.1 African Peer Review Mechanism African Governance Report 2019

In 2017, at the 28th Ordinary session of the AU Assembly of Heads of State and Government, the APRM mandate was increased to include monitoring the implementation and progress towards Agenda 2063 and the SDGs outlined in Agenda 2030. Thereafter, the African Governance Architecture (AGA) Platform developed a work plan in this regard and tasked the APRM with preparing the Africa Governance Report (AGR). The AGA Platform comprises various AU organs, institutions and RECs and facilitates dialogue targeted at "the harmonization and coordination of instruments and initiatives for promoting good governance, democracy, the rule of law, and human rights",[149] with a specific focus on the AU shared values.[150]

---

[148]Ibid.

[149]African Peer Review Mechanism, *The African Governance Report—Promoting African Union Shared Values*, South Africa, 2019.

[150]Ibid.

The inaugural AGR was released in January 2019 and forms a baseline for tracking future progress and evaluating national, regional and continental trends.[151] The AGR investigates the implementation of the AU shared values in five specific areas:

1. Transformational leadership,
2. Constitutionalism and the rule of law,
3. Peace, security and governance,
4. Nexus of development and governance and
5. The role of RECs in African governance.

These five core areas of governance relate closely with four of the Agenda 2063 Aspirations, namely Aspiration 1, 3, 4 and 6. While numerous SDGs from Agenda 2030 overlap with the key governance areas, the AGR notes that SDG 16 ("Promote peaceful and inclusive societies for sustainable development, provide access to justice for all and build effective, accountable and inclusive institutions at all levels"[152]), in particular, is well-aligned with its focus areas. The APRM's AGR will serve to inform the general public as well as member states, RECs and other AU instruments on the trends and movements of governance on the African continent.

While it is true that various assessments and studies examining governance throughout Africa are readily available, the AU noted that an African-generated governance report is in line with the AU decision to be in control of its own account-ability and monitoring. Furthermore, the fact that the APRM is an AU organ allows it "unfettered access to member state informants and state-held data". Lastly, the APRM highlights the fact that since the report and its recommendations are devel-oped by "Africans for Africa", it "improves prospects for the implementation of its recommendations".[153]

The research and analysis process comprised discussions with representatives from AU member states, AU organs, RECs and other stakeholders from both civil society institutions and the private sector, selective surveys and analysis of existing reputable data on governance indicators, socio-economic studies, and peace and security reports. The AGR notes that governance throughout Africa has improved, with the largest gains observed in socio-economic development.[154] The least progress was found to be in democracy and political governance, which is supported by the findings in both the Freedom House report and the MIF African Governance Report. A summary of the findings for each of the five key areas is offered below[155]:

1. **Transformative Leadership**: Transformative leadership at all levels (conti-nental, regional, nation and sub-national) is paramount to realizing the AU goals by galvanizing the practice of democracy and driving progressive change. Africa

---

[151] Ibid.

[152] United Nations, Sustainable Development Goals Knowledge Platform, Sustainable Development Goals, https://sustainabledevelopment.un.org/sdgs, accessed: 8 February 2020.

[153] African Peer Review Mechanism, *The African Governance Report—Promoting African Union Shared Values*, South Africa, 2019.

[154] Ibid.

[155] Ibid.

has exhibited good progress towards protecting democracy and formulating National Visions, but more work needs to be done in the following areas[156]:

(a) Aligning national plans with Agenda 2063 and the UN SDGs,
(b) Ensuring prejudice, discrimination and exclusion are eliminated,
(c) Promoting good governance principles, ensuring public participation and protecting democracy and
(d) Ensuring the protection of human rights.

2. **Constitutionalism and The Rule of Law**: African member states, in general, have demonstrated progress in specific areas relating to constitutionalism and the rule of law, particularly in: encouraging democratization, respecting the time frames of presidential terms, upholding human rights, and introducing new methods for monitoring institutions. The AGR noted that progress is required in certain areas, such as[157]:

(a) Developing better anti-corruption programmes and strategies,
(b) Revising policies relating to criminal justice systems to better incorporate customary practices and
(c) Ensuring member states uphold the reporting requirements of the various AU instruments.

3. **Peace, Security and Governance**: There are numerous AU and international instruments centred on ensuring peace throughout Africa, notably the AU "Silencing the Guns" by 2020 initiative. Overall international wars declined in Africa, however, violent conflicts, intra-state uprisings and terrorism have increased over the last 20 years. Recommendations in this regard include[158]:

(a) Ensuring all relevant AU instruments are signed, ratified and domesticated by member states,
(b) Improving coordination between the AU and RECs in relation to promoting and ensuring peace and security within the continent, and
(c) Increasing the capacity and deployment capacity of the Africa Standby Force.

4. **Nexus of Development and Governance**: The national visions and development strategies of many member states are not adequately aligned with AU Agenda 2063 and the UN SDGs and many face implementation challenges. The AGR recommends[159]:

(a) Ensuring National Development Plans, National Visions and National Action Plans are aligned with Agenda 2063 and the UN SDGs (possibly through the APRM Country Review reports),

---

[156]African Peer Review Mechanism, *The African Governance Report—Promoting African Union Shared Values*, South Africa, 2019.
[157]Ibid.
[158]Ibid.
[159]Ibid.

(b) Ensuring inclusive human development is at the core of all development plans and

(c) Ensuring realistic, achievable plans focused on sustainable utilization of resources.

5. **Role of the RECs in African Governance**: RECs focus on "continental unity, development, economic cooperation and integration, and promotion of democracy and peace and security". However, better synchronization of their functions and responsibilities with those of the AU is required. The AGR proposes[160]:

(a) Enhancing collaboration between the RECs and the AU, and

(b) Ensuring the development plans of the RECs are well-aligned with the Agenda 2063 and UN SDGs.

The APRM AGR recommendations generally focus on promoting the ratification of AU instruments by all member states. Furthermore, the report highlights that member states must amend their National Visions, development plans and policies to ensure they align with the AU Agenda 2063 and the UN SDGs.

1.5.2 Mo Ibrahim Foundation—Ibrahim Index of African Governance

The MIF was established in 2006 to define, assess and enhance governance levels across the African continent through four core initiatives, namely:

1. The Ibrahim Index of African Governance (IIAG),
2. The Ibrahim Leadership Prize,
3. Ibrahim Governance Weekend and
4. The Ibrahim Fellowships programme focused on mentoring future African leaders.

The MIF defines governance as: "the provision of the political, social and economic public goods and services that every citizen has the right to expect from their state, and that a state has the responsibility to deliver to its citizens".[161]

The IIAG is a comprehensive instrument for measuring governance performance. The African Governance Report is issued annually and assesses each African country and determines its Overall Governance performance ranking. Each country's governance performance is assessed according to four central categories, particularly: (i) safety and rule of law, (ii) participation and human rights, (iii) sustainable economic

---

[160]Ibid.

[161]Mo Ibrahim Foundation, *Ibrahim Index of African Governance (IIAG)*, https://mo.ibrahim.fou ndation/iiag, accessed: 8 February 2020.

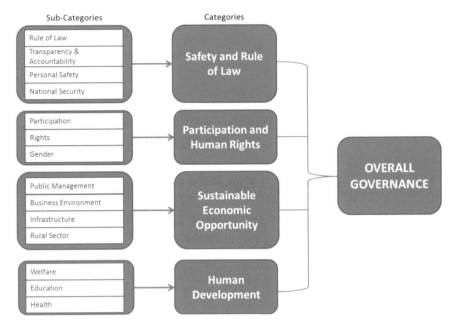

Fig. 1.14 Ibrahim Index of African Governance categories and subcategories (Mo Ibrahim Foundation, *Ibrahim Index of African Governance (IIAG)*, https://mo.ibrahim.foundation/iiag, accessed: 8 February 2020)

opportunity, and (iv) human development. Each of these categories contains subcategories comprised of various performance indicators in order to quantify the overreaching success of governance, as shown in Fig. 1.14. There is a total of 14 subcategories, comprising 102 indicators.

The IIAG document is based on data from over 35 different sources (including official data sets, expert assessment reports and perception surveys) across 191 different variables. In the latest 2018 report, the collected data and additional calculations performed for the IIAG resulted in a total of 273 different measures of governance for any African state for any given year between 2008 and 2017.[162] As such, the IIAG represents the most comprehensive tool to assess a country's progress towards both Agenda 2063 and the 2030 SDGs. Good governance is a key requirement in realizing the goals set out in both agendas. In this vein, the SDG Center for Africa states "while good governance is a Sustainable Development Goal in its own right (SDG16), the active role of government is instrumental to every other SDG as well. For this reason, it is necessary not only for each country to reflect on its current SDG

---

[162]Mo Ibrahim Foundation, *African Governance Report*, 2019.

status and progress (…) but to analyse how government-led efforts can be improved and accelerated."[163]

The governance measures in the IIAG cover six of the seven Agenda 2063 aspirations and 70% of its goals. Moreover, these governance measures cover twelve of the 17 UN SDGs.[164]

Key findings from the 2019 IIAG African Governance Report are summarised below[165]:

- Following the implementation of the FTYIP, access to education across Africa has increased, however, the average continental score for education has deteriorated due to a decrease in the quality of education within the continent. This is exacerbated by an increasing mismatch between job market requirements and African education systems.
- Similarly, the African average for the *Promotion of Socio-economic Integration of Youth* indicator has shown an annual decline since 2014.
- Basic nourishment has, on average, shown a continual decline over the same time period.
- Human trafficking indicators show an increase, on average, across the continent since 2014. Moreover, African countries must do more to strengthen their laws addressing gender-based violence.
- Unconstitutional regime changes are still a reality in Africa, but the AU Agenda 2063 specifies zero tolerance in this regard.
- The indicator for citizen's dissatisfaction with *Basic Health Services* has increased since 2009. The report also notes that on average, in Africa there are only 17 skilled health workers per 10,000 citizens (compared to 117 per 10,000 in the United States of America (USA)).
- *Transparency and Accountability* scores remain the lowest out of all the 14 subcategories and the indicator for *Absence of Corruption in the Private Sector* has decreased significantly, highlighting how corruption is inhibiting governance and development throughout the continent.
- Indicators relating to critical infrastructure within African countries still remain low.
- The lowest scoring of all 102 indicators was *Diversification of Exports*.
- NSOs in Africa are suffering from limited autonomy and poor funding, and 45.6% of the African population lives in countries where a census has not been performed in the past decade.
- Vital statistics and data coverage for population and education have shown continued deterioration since 2015, with only eight African countries achieving more than 90% coverage rates for birth registrations and only three countries having a death registration coverage rate in excess of 90%.

---

[163]The Sustainable Development Goals Center for Africa, *2019 Africa—SDG Index and Dashboard Report*, June 2019.

[164]Mo Ibrahim Foundation, *African Governance Report*, 2019.

[165]Ibid.

The IIAG African Governance Report determined that more than 80% of the Agenda 2063 targets (255 in total) do not have well-defined core indicators that can be utilized for monitoring progress.[166] This must be addressed by the AU to ensure better monitoring of progress in the next TYIP.

Likewise, monitoring progress towards realizing the UN SDGs, the study revealed that:

- There is no internationally accepted standards or methodology available for monitoring almost 20% of the SDG indicators.
- The UN database tool for tracking progress towards the SDGs only has data for 91 out of the 232 SDG indicators for African countries.
- Due to a lack of meaningful data, more than 50% of data sources used for calculating SDG indicators come from estimation, global monitoring or modelling when it comes to Africa.[167]

1.6 Conclusion

Good governance is crucial in assisting countries to realize the UN SDGs and the Africa 2063 Agenda. It has been demonstrated that evaluating the progress achieved by countries towards the two agendas offers insights into the governance efficacy within the countries. More African countries must align their national development strategies with the Africa 2063 Agenda to achieve increased progress.

This chapter has revealed that continued implementation of governance requires a complex framework of political institutions supported by administrative bodies at a continental, regional and national level. The continental frameworks, specifically, the AU, the New Partnership for Africa's Development (NEPAD), the African Peer Review Mechanism (APRM) and the African Continental Free Trade Area (AfCFTA), were examined in detail. The eight RECs comprising the AU, are critical in ensuring good governance within Africa, however, better synchronization with the AU is required to maximize their impact.

This chapter has investigated the four fundamental challenges prohibiting good governance within Africa, namely corruption, weak democratic institutions, limited data coverage and openness, and a lack of civil registration and vital statistics. Analysis revealed corruption is undermining democratic processes and weakening governance in African countries. Examination of freedom levels across the continent exhibited numerous authoritarian governments and an increase in repression of dissent. Statistical capacity and meaningful, contemporary data availability within the continent remain a major hurdle facing good governance, with numerous African countries exhibiting a lack of governmental statistical capacity, inadequate funding of national statistics offices and a lack of independence of such offices. Thereafter, it was demonstrated how governance within certain African states is crippled by a

---

[166]Ibid.
[167]Ibid.

lack of civil registration and vital statistics, with many African states exhibiting poor birth and death registration coverage data and an absence of effective identification systems.

Lastly, the current governance levels in Africa were established through an analysis of the latest versions of the APRM African Governance Report and the IIAG African Governance Report. Analysis confirmed that although governance throughout Africa has improved, the least progress was achieved in democracy and political governance, with unconstitutional regime changes still being a reality on the continent. Moreover, it was confirmed that corruption, a lack of accountability and transparency and poor data availability continue to challenge African governance and development.

A substantial lack of well-defined indicators for monitoring Africa's progress towards Agenda 2063 was identified. Similarly, it was shown that a lack of meaningful governance data within Africa contributes to the difficulties in assessing its progress towards attaining the UN 2030 SDGs.

Chapter 2
E-Governance in Africa and the World

Abstract This chapter focuses on e-government and e-governance by assessing the e-government readiness levels across Africa through an analysis of the key findings of the United Nations E-Government Survey from 2018. New trends and emerging technologies in e-government including blockchain, e-voting and digital government are discussed thereafter. Specific case studies of innovative e-government applications from both developed and developing countries are examined to identify objectives and strategies that African countries should consider when implementing their own e-governments.

2.1 Introduction

No man is good enough to govern another man without the other's consent
Abraham Lincoln

This chapter examines the concept of e-governance and e-government in detail and investigates the definitions thereof. It explores the prerequisites for the successful establishment of e-governance systems as well as current implementation and adoption strategies. It discusses the UN *2018 E-Governance Survey*, which assesses e-governance levels across all 193 member states, with a specific focus on African countries and trends. A selection of exemplary e-government solutions that highlight the potential of e-governance are discussed thereafter.

Following the 15th Meeting of the Group Experts on the UN Programme in Public Administration and Finance, it was decided that the UN Department of Economic and Social Affairs (UN DESA) would prepare and publish the *World Public Sector Report* every two years, assessing "major trends and issues concerning public administration and governance".[1] The first report, published in 2001, focused on globalization and its effect on states and their governance practices.[2]

---

[1] UN DESA, World Public Sector Report: Globalization and the State, 2001.
[2] Ibid.

A. Froehlich et al., *Space Supporting Africa*, Studies in Space Policy 28,
https://doi.org/10.1007/978-3-030-52260-5_2

53

The second *World Public Sector Report*, published in 2003, titled "E-Governance at the Crossroads", identified the importance of information and communications technology (ICT) infrastructure in aiding and assisting governments with operational cost reduction, increasing citizen participation and service delivery, and performance optimization. The report discussed methods and frameworks to ensure e-government applications were both meaningful and successful after the UN received numerous requests from member states for advice on integrating ICT within their government operations.

In this regard, UN DESA defined e-government as

"a government that applies ICT to transform its internal and external relationships"and noted that "through the application of ICT to its operations, a government does not alter its functions or its obligation to remain useful, legitimate, transparent and accountable".[3]

As noted in Chap. 1, the EU defines e-government as

"the use of information and communication technology in public administrations combined with organisational change and new skills in order to improve public services and democratic processes and strengthen support to public policies".[4]

The UN 2030 Agenda for Sustainable Development recognized the vast potential for e-government to assist in realizing the 2030 SDGs and promote global development. It states, "the spread of information and communications technology and global interconnectedness have great potential to accelerate human progress to bridge the digital divide and to develop knowledge societies, as does scientific and technological innovation across areas as diverse as medicine and energy".[5]

2.2 United Nations *2018 E-Governance Report*

The 2018 *UN E-Government Survey* has been published by UN DESA since 2001 and is now on its tenth edition. The survey assesses the progress and status of e-government development and effectiveness in delivery of public services across all 193 member states of the UN and provides useful evidence and insights highlighting the potential of e-governance to facilitate realizing Agenda 2030.[6]

The collection and assessment of data relating to e-government development in member states is based on a holistic view that covers three core dimensions: i) the level of ICT infrastructure of a state, ii) the state's ability and human capacity to promote and use ICTs, and iii) the availability of online government information

---

[3] UN DESA, World Public Sector Report: E-Governance at the Crossroads, 2003.

[4] Ibid., p. 25.

[5] United Nations, Transforming our World: the 2030 Agenda for Sustainable Development, 2015, A/RES/70/1, para. 15.

[6] United Nations Department of Economic and Social Affairs, *E-Government Survey 2018—Gearing E-Government to Support Transformation towards Sustainable and Resilient Societies*, New York, 2018.

and services. Data are collected using a questionnaire issued to member states and are supplemented with in-depth literature reviews and input from both scholars and practitioners.

The data are analysed as per a well-defined framework that provides a single index, labelled the E-Government Development Index (EGDI), which is determined using the weighted average of three normalized indices, namely: the Telecommunications Infrastructure Index (TII), the Human Capital Index (HCI), and the Online Service Index (OSI). The TII is calculated using International Telecommunications Union (ITU) data, while data provided by the UN Educational, Scientific and Cultural Organization (UNESCO) is employed to determine the HCI. The OSI is calculated using data collected from an independent survey questionnaire that assesses the national online presence of states. This includes evaluating the availability of online and mobile service delivery and open government data as well as e-participation, citizen engagement/usage, digital divides and innovative ICT initiatives within the state. The OSI measures a government's use of ICTs to deliver services at a national level. This methodological framework has remained consistent since 2001, with minor adjustments (such as the inclusion of new indicators) made as necessary to reflect new trends and technologies in e-government. The EGDI measures "the readiness and capacity of national institutions to use ICTs to deliver public services".[7]

The theme of the 2018 E-Government Survey is "gearing e-government to support transformation towards sustainable and resilient societies" to align with the focus of the 2018 High-level Political Forum (HLPF),[8] which speaks directly to SDG 11—"make cities and human settlements inclusive, safe, resilient and sustainable".[9] Ensuring resilient societies underpins the continued realization of all the SDGs. The 2018 Survey investigates the different technologies and methods governments use to increase their response to natural and human-made disasters. With this theme in mind, the 2018 Survey report assesses the following:

- mobilizing e-government to build resilient societies,
- inclusivity of e-government by overcoming digital divides and other challenges,
- global and regional perspectives on building e-resilience through e-government,
- the resilience of e-governments to cybersecurity threats,
- leveraging e-government to improve the resilience and sustainability of cities, and
- new technologies in e-government, including artificial intelligence (AI) and big data.

This section initially discusses the latest global trends in e-governance and thereafter assesses Africa's performance in the 2018 Survey and concludes with an analysis of key thematic areas relevant to developing countries seeking to implement e-government solutions.

---

[7]Ibid.

[8]Ibid.

[9]United Nations, Sustainable Development Goals Knowledge Platform, *Sustainable Development Goal 11*, https://sustainabledevelopment.un.org/sdg11, accessed: 21 February 2020.

The UN DESA *E-Government Survey* Report is a salient tool for assessing progress towards the attainment of SDG 16, which is to "promote peaceful and inclusive societies for sustainable development, provide access to justice for all and build effective, accountable and inclusive institutions at all levels."[10] It highlights the progress public institutions have made across five key dimensions, namely: effectiveness, inclusion, openness, trustworthiness and accountability.[11]

2.2.1 Global Trends and Africa's Performance in the 2018 UN E-Government Survey

Countries are rated on their EGDI performance according to the following classifications:

- Very High: EGDI Score between 0.75 and 1.00 (1.00 being the highest score)
- High: EGDI Score between 0.50 and 0.75
- Middle: EGDI Score between 0.25 and 0.50
- Low: EGDI Score less than 0.25

The 2018 report shows a consistent positive global trend, and increased e-government uptake and readiness levels throughout the world since 2001. In 2018, a total of 40 countries were ranked as Very High—a number that has quadrupled in comparison with 2003 results. Furthermore, since 2014, all 193 member states have realized some form of online presence. Moreover, almost two thirds of them fall within the High and Very High categories. Currently the most significant barriers facing e-governance adoption are:

(1) ensuring data security and privacy,
(2) cybersecurity and resilience,
(3) ensuring inclusive e-governance by overcoming digital divides and digital literacy challenges, and
(4) increasing e-participation.[12]

In general, the 2018 results reveal Africa's consistent underperformance in the e-government survey. Figure 2.1 depicts the EGDI performance of African countries in the 2018 survey and highlights that no African countries placed in the Very High category. However; it is worth noting that 38 of the 40 countries in the Very High category are high-income countries.

---

[10]United Nations, Sustainable Development Goals Knowledge Platform, *Sustainable Development Goal 16*, https://sustainabledevelopment.un.org/sdg16, accessed: 21 February 2020.

[11]Ibid.

[12]United Nations Department of Economic and Social Affairs, *E-Government Survey 2018— Gearing E-Government to Support Transformation towards Sustainable and Resilient Societies*, New York, 2018.

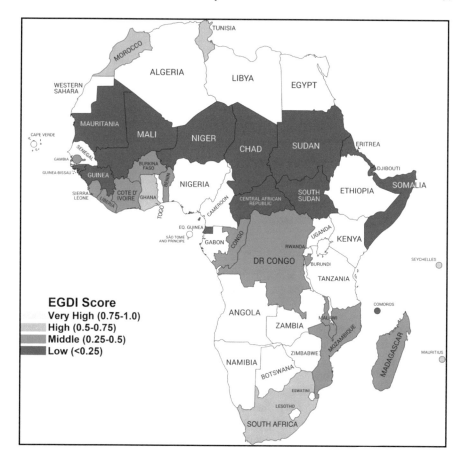

Fig. 2.1 EGDI performance of African countries in the 2018 UN E-Government Survey (Ibid.)

Six African countries fell in the High category, specifically Ghana (that transitioned from the Middle category after considerable development of ICTs under the e-Ghana and e-Transform projects, which are discussed in detail later in this chapter), Mauritius, Morocco, Seychelles, South Africa and Tunisia.[13] Africa's poor performance in the E-Governance Survey is depicted in Fig. 2.2, which shows the 2018 regional average EGDI scores.

Only four African countries achieved EGDI scores higher than the world average of 0.55, namely Mauritius, South Africa, Tunisia and Seychelles. These are the only four African countries that feature in the top 100 global EGDI rankings.[14]

Figure 2.3 shows the regional composition of the four different EGDI classifica-

---

[13]Note that the countries are listed alphabetically and not according to their performance in the E-Government Survey. Table 2.3 depicts the top ten African countries as per the survey results, as well as the scores they achieved for EGDI, OSI, HCI and TII.

[14]Ibid.

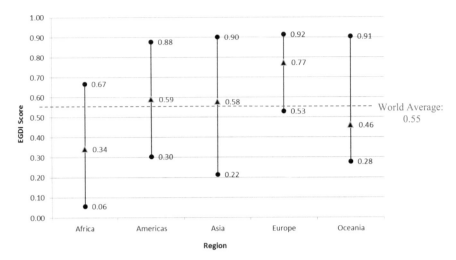

Fig. 2.2 Regional average EGDI values (including maximum and minimum) for 2018 (Image recreated from data contained within the UN DESA *E-Government Survey* 2018, p. 92)

tions, specifically Low, Middle, High and Very High. Africa's poor performance is highlighted in the regional breakdown—African countries account for 87% of the Low level group, 50% of the Middle group, 9% of the High level group and 0% of the Very High level group.

The number of countries in the Middle category in the 2018 report is 66 with 18 of these countries transitioning from the Low category in 2016. This reveals significant global e-government development. Twelve of the countries that moved up into the Middle classification are African, specifically Benin, Burkina Faso, Burundi, Congo, Democratic Republic of Congo, Gambia, Ivory Coast, Liberia, Madagascar, Malawi, Mozambique and Sierra Leone.[15]

Movements within the Low EGDI category confirm the global move towards better e-government. In 2016, 32 countries were placed within this category and in 2018 this reduced to just 16 countries. 14 of the Low EGDI countries are African and represent some of the least developed countries in the world. The African countries in the Low category are the Central African Republic, Chad, Comoros, Djibouti, Equatorial Guinea, Eritrea, Guinea, Guinea Bissau, Mali, Mauritania, Niger, Somalia, South Sudan and Sudan (which fell from a Middle classification in 2016 due to adverse political activity).[16]

The global trend towards better e-government can also be observed within Africa, however at a slower rate. Figure 2.4 depicts the EGDI classifications of African countries in the UN E-Government Survey in 2014 and 2018. The number of African

---

[15]United Nations Department of Economic and Social Affairs, *E-Government Survey 2018—Gearing E-Government to Support Transformation towards Sustainable and Resilient Societies*, New York, 2018.

[16]Ibid.

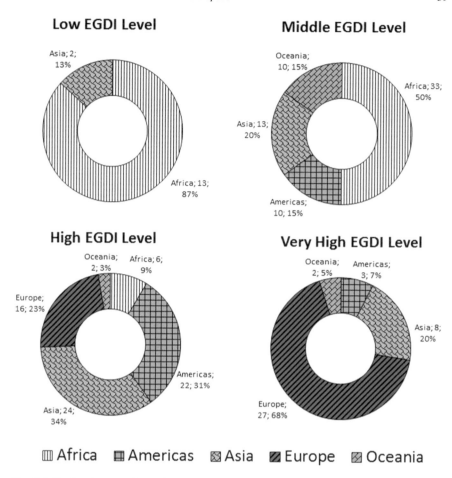

Fig. 2.3 Regional composition of EGDI level groups, 2018 (Image created from data contained within the UN DESA *E-Government Survey 2018*)

countries in the Low classification halved over the four year period from 26 countries to 13 countries. The number of African Middle level countries increased from 23 to 33 countries, and the High level countries also increased from five in 2014 to eight countries in 2018. Unfortunately, no African countries transitioned to the Very High classification; however, the large number of countries leaving the Low classification (representing 24% of the continent's countries) shows there is a continent-wide shift towards better e-governance.

Global trends relating to e-governance that were identified in the 2018 report are listed and summarised below[17]:

[17]United Nations Department of Economic and Social Affairs, *E-Government Survey 2018— Gearing E-Government to Support Transformation towards Sustainable and Resilient Societies*, New York, 2018.

2014 EGDI Classifications

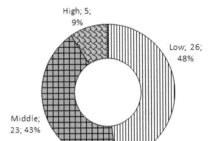

2018 EGDI Classifications

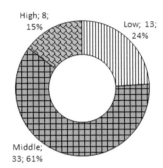

▥ Low ⊞ Middle ⊠ High ▨ Very High

Fig. 2.4 Comparison of EGDI classifications of African countries between 2014 and 2018 (Figure created from data within the 2014 and 2018 UN DESA *E-Government Survey Report*)

1. Online Service Delivery and OSI scores:

   (a) Progress in OSI scores generally correlate with improvements in EGDI scores.
   (b) Countries with OSI scores higher than their overall EGDI scores generally exhibit slower development in telecommunications infrastructure and human capital.
   (c) All regions globally exhibited an improvement in online services.

2. All 193 member states have national portals, however not all of them provide transactional online services. The three most common services provided by governments online include paying for utilities (140 countries), submitting income taxes (139 countries) and registering new businesses (126 countries). Countries providing online birth registrations increased from 44 in 2016 to 86 in 2018, showing significant improvement. However, this only represents 47% of the total UN member states. The number of countries providing other online transactional services in 2018 is indicated in Fig. 2.5. Additional new online transactional services identified by the 2018 report and the number of member states offering these services are shown in Table 2.1.

3. Digital technologies including emails, short message service (SMS) updates, rich site summary (RSS) feeds, mobile applications and downloadable forms are being used to increase the number of online services across all sectors. Services provided through mobile applications have shown the fastest growth at 52% in the education, employment and environment sectors. But Africa still falls well behind the global averages in all sectors, as highlighted in Fig. 2.6.

4. Globally, e-government is leading to increased inclusivity:

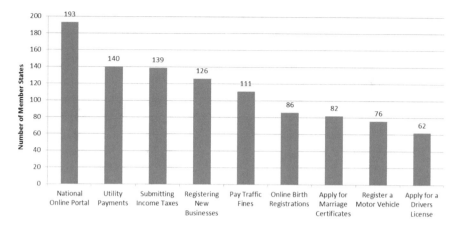

Fig. 2.5 Number of countries offering online transactional services in 2018 (Image created from data contained within the UN DESA *E-Government Survey 2018*)

Table 2.1 Number of countries offering new online transactional services Identified in the 2018 UN DESA *E-Government Survey*

Online transactional service description	Number of member states offering service
Apply for government vacancies	129
Submit Value Added Tax (VAT)	121
Apply for a business license	104
Apply for visas	100
Apply for social protection programs	91
Make police declarations	84
Apply for death certificates	78
Apply for land title registration	67
Register changes of address	61
Apply for building permits	58

Table created from data contained within the UN DESA *E-Government Survey 2018*

(a)   The number of countries providing targeted online transactional services aimed at vulnerable or minority groups is shown in Fig. 2.7. Compared with the 2016 survey results, the number of countries with online services for low-income groups has almost tripled (120 countries in 2018), and the number of countries providing online services specifically for youth, women, immigrants, elderly persons, and persons with disabilities has almost doubled.

(b)   The number of African countries offering specific online transactions for vulnerable or minority groups is displayed in Fig. 2.8. The percentage of

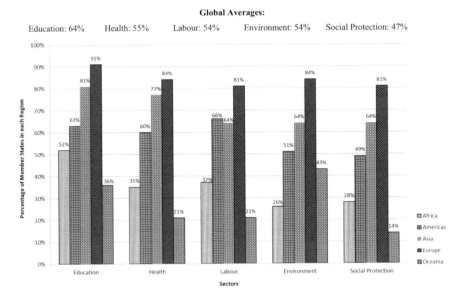

Fig. 2.6 Percentage of member states in each region which provide services via e-mail, SMS or RSS in different sectors in 2018 (Image recreated from data contained within the UN DESA *E-Government Survey* 2018, p. 102)

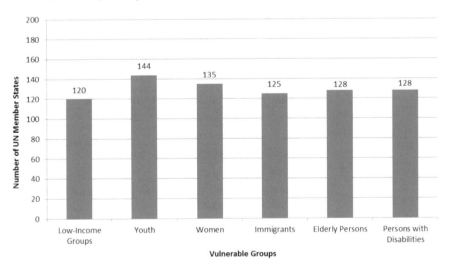

Fig. 2.7 Number of member states offering targeted online services for vulnerable/minority groups (Image created from data contained within the UN DESA *E-Government Survey* 2018)

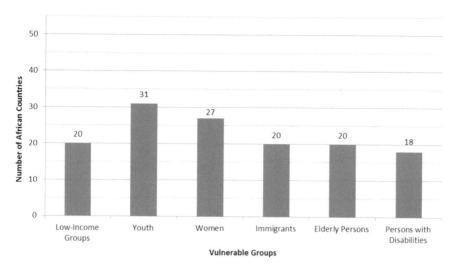

Fig. 2.8 Number of African countries offering targeted online services for vulnerable/minority groups (Image created from data contained within the UN DESA *E-Government Survey* 2018)

African countries offering online service delivery for vulnerable groups rose from 33% in 2016 to 57% in 2018.

5. There is a noted increase in transparency, accountability and openness in governance as a result of e-government uptake:

(a) The number of countries with online public mechanisms relating to e-procurement, public bidding and tender adjudication processes has increased—160 countries have online announcements of public tenders, 130 have an online e-procurement platform, 149 countries publish bidding results online, and 115 countries provide information for monitoring and evaluating public procurements online.

(b) More countries globally are listing government vacancies and employment opportunities within the public sector online, which results in increased transparency in the recruitment process. In 2018, 25 African countries were found to announce government vacancies online, highlighting a significant increase from just nine countries in 2016.

(c) The number of countries with open government data (OGD) portals where government datasets (including data from the education, health, environment, social programmes and finance departments) are shared online in open formats increased from 46 countries in 2014 to 139 in 2018. Furthermore, OGD portals are exhibiting increased functionality, better metadata and, at times, even dedicated mobile applications. This increases transparency and accountability in the governance process and ultimately contributes to improving trust in public institutions.

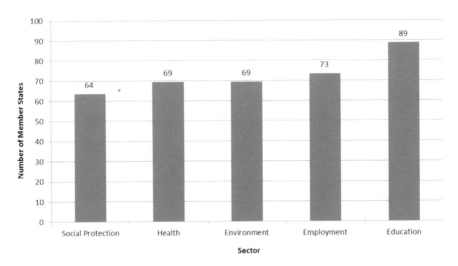

Fig. 2.9 Number of member states using mobile applications and SMS services in each sector in 2018 (Image created from data contained within the UN DESA *E-Government Survey* 2018)

6. There has been a global increase in the application of mobile networks and devices to realize e-governance across all sectors. The most noticeable of these is in the education sector, which increased from 30% of countries in 2016 to 46% in 2018 (which represents 89 countries). Figure 2.9 portrays the number of member states currently utilizing mobile services and/or applications in the various sectors.

7. E-participation is a crucial element of e-governance and is defined by the UN as "the process of engaging citizens through ICTs in policy, decision-making, and service design and delivery so as to make it participatory, inclusive, and deliberative". E-participation is measured in the 2018 Survey using the E-Participation Index (EPI), which is derived from three scores, namely: e-information (the amount of government information available online), e-consultations (online forums and platforms facilitating public consultations and engagement) and e-decision-making (which involves citizens in the decision-making process using online tools). Noteworthy trends and findings from the 2018 Survey relating to the EPI are listed below:

   (a) The number of countries with Very High EPI ratings doubled from 31 in 2016, to 62 in 2018. The five best performing countries were Denmark (where e-participation is included in the country's Digital Strategy), Finland, South Korea, the Netherlands and Australia (where legislation mandates all new public-facing services must be accessible to all users).

   (b) Four of the Very High rated countries are in Africa, namely: Morocco, Rwanda, South Africa and Tunisia. Noticeable movements in African countries include Burkina Faso (which moved up 56 places in the EPI rankings since 2016), Central African Republic (which moved up 40 places), Djibouti and Sierra Leone who both increased their position on the EPI rankings by 38 positions. This indicates a positive shift within African governments to increase their commitment to improving e-participation and encouraging more inclusive governance styles.

(c)   Facilitating e-consultation requires online engagements tools that allow citizens to provide inputs on new policies, legislation, services and/or projects. The number of African countries with such engagement tools (as of 2018) is listed below. It is, however, important to note that the presence of such tools does not necessarily guarantee the received inputs are duly considered by governments:

- Social media engagement tools: 46 countries
- National portals with e-tools for public consultation: 24 countries
- Online consultation e-tools for development: 46 countries
- No online tools at all: 2 countries.

(d)   E-Decision making, the third pillar of e-participation, continues to have the lowest global uptake and remains a serious challenge.

8.   Public-private partnerships (PPPs) as a vehicle for implementing aspects of e-government have emerged as one of the most successful recent models for the provision of public services online, specifically in the education, health, environmental and financial sectors. Africa has shown the highest regional increase in the number of countries providing online services through civil society or private sector partnerships, with a total number of 44 countries in 2018 (an increase from 23 African countries in 2016). While PPPs allow for vertical integration of services, shared risk profiles and increased resources, they require detailed and methodical contracts with defined risk sharing and performance indicators to ensure successful implementation.

Comparing Africa's EGDI average scores with the other regions in the world (see Table 2.2), it is evident that Africa is the worst performing region in both overall EGDI scores, as well as in all its sub-indices: Online Service Index (OSI), Human Capital Index (HCI), and Telecom Infrastructure Index (TII).

Table 2.2   Regional EGDI, OSI, HCI and TII scores

Region	Average EGDI score	Average OSI score	Highest regional HCI score	Lowest regional TII score
Africa	0.34	0.36	0.46	0.20
Americas	0.59	0.61	0.72	0.44
Asia	0.58	0.62	0.67	0.44
Europe	0.77	0.80	0.85	0.68
Oceania	0.46	0.39	0.71	0.28

Ibid., p. 128

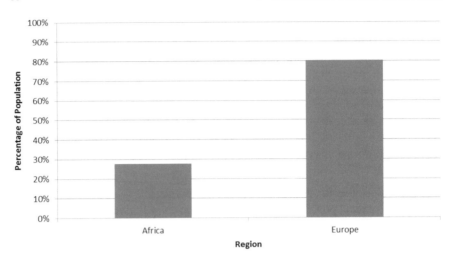

Fig. 2.10 Comparing the percentage of population with mobile subscriptions—Africa versus Europe (Image created from data contained within the UN DESA *E-Government Survey* 2018)

Africa exhibits the least developed technical and ICT infrastructure in the world. This, coupled with poor Internet connectivity in rural areas, contributes to its weak performance in terms of e-government readiness. Access to fixed-line broadband connections is extremely limited across the continent, with many African countries having no fixed-line infrastructure at all. The 2018 trends revealed that the number of fixed broadband subscriptions per 100 people in Africa was just 2.16, which increased from 1.2 in 2016.[18]

Mobile broadband access has increased significantly across the continent, but the high relative cost of access and low subscription levels still make Internet access a significant barrier to e-government systems in Africa. The severity of these issues becomes apparent when comparing African figures to European ones—while 80.46% of the population in Europe has mobile subscriptions, Africa, in comparison, has only 27.84% as demonstrated in Fig. 2.10.[19]

Furthermore, the price of mobile broadband access is comparatively very high in Africa, costing 13.49% of its gross national income per capita to access mobile broadband, compared with 0.63% in Europe (as highlighted in Fig. 2.11). This relatively high cost is a barrier that results in exclusion of people who cannot afford such costs.[20]

The 2018 Survey reveals the commitment of many African states to increasing their OSI presence, which in turn has led to an overall increase in the EGDI ratings within the region. The top ten performing countries in Africa, in terms of

[18]Ibid.

[19]Ibid.

[20]United Nations Department of Economic and Social Affairs, *E-Government Survey 2018—Gearing E-Government to Support Transformation towards Sustainable and Resilient Societies*, New York, 2018.

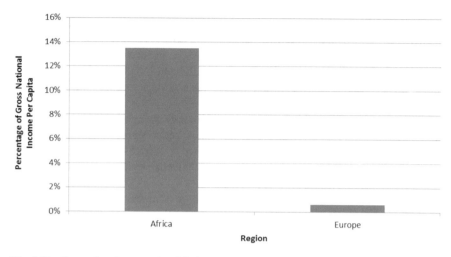

Fig. 2.11 Comparing the cost of mobile broadband access in Africa and Europe (Image created from data contained within the UN DESA *E-Government Survey* 2018)

e-government readiness, are shown in Table 2.3 alongside their 2018 EGDI, OSI, HCI and TII scores.

While it could be assumed that increasing the online availability of public services automatically translates to improving e-government within a nation, the low TII scores for all of the top ten African countries reiterates one of the major challenges to e-government in Africa, specifically: the lack of ICT infrastructure across the continent.

Table 2.3 Top ten African countries in the 2018 UN *E-Government Survey*

2018 Rank in Africa	Country	EGDI score	2018 global rank	OSI score	HCI score	TII score
1	Mauritius	0.668	66	0.729	0.731	0.544
2	South Africa	0.662	68	0.833	0.729	0.423
3	Tunisia	0.625	80	0.806	0.664	0.407
4	Seychelles	0.616	83	0.618	0.730	0.501
5	Ghana	0.539	101	0.694	0.567	0.356
6	Morocco	0.521	110	0.667	0.528	0.370
7	Cape Verde	0.498	112	0.486	0.615	0.393
8	Egypt	0.488	114	0.535	0.607	0.322
9	Rwanda	0.459	120	0.722	0.482	0.173
10	Namibia	0.455	121	0.451	0.585	0.330

Ibid., p. 135

This, combined with the improved OSI scores of African countries, actually represents a risk—with increased online presence but poor ICT infrastructure, the likelihood that certain groups of the population will be excluded from e-government rises. This is discussed in more detail in Sect. 2.2.3 and is a concern which must be addressed to ensure inclusive, fair and ubiquitous e-government solutions in African countries.

The HCI scores of the top ten African countries indicate a commitment to capacity building and human capital development in e-government applications. However, this is not representative of the average scores achieved by Africa as a whole, where low human capital development is another major challenge to the implementation of e-government.

UNECA is actively involved in advancing e-government in Africa and assists countries in adopting evidence-based policies to promote ICT development and promoting collaboration with different stakeholders, both locally and internationally, in the Internet community. UNECA launched the African Information Society Initiative (AISI), which resulted in 48 African countries establishing national e-strategies aligned with Agenda 2030 and has been researching emerging technologies in the digital economy and how these can be implemented in the African context and leveraged to maximize their socio-economic benefits. UNECA also established the Partnership on Measuring Information and Communications Technology for Development in 2004, which aims to improve the standardization and collection of internationally comparable ICT statistics and development indicators.[21]

The UNECA and the AUC drafted the AU Convention of Cyber Security and Personal Data Protection, which was adopted at the 23rd Ordinary Session of the Assembly in 2014. However, as of June 2019, the convention had only been signed by 14 African countries and ratified by the following five: Ghana, Guinea, Mauritius, Namibia and Senegal.[22] The Convention is discussed in greater detail in Sect. 2.2.4 where cyber security issues relating to e-government applications are addressed.

2.2.2   Prerequisites for E-Government Implementation

The 2018 Survey report states that "deploying e-government in support of good governance is essential for building effective, accountable and inclusive institutions

---

[21] Ibid.

[22] African Union, *List of countries which have signed, ratified/acceded to the African Union Convention on Cyber Security and Personal Data Protection*, Ethiopia, June 2019.

at all levels, as called for in SDG 16,[23] and for strengthening implementation of SDG 17,[24] both of which underpin achievement of the SDGs as a whole".[25]

To ensure the successful implementation of e-government applications and systems, certain key prerequisites must be satisfied. The prerequisites identified by UN DESA in the 2018 Survey report are discussed in this section, which also identifies specific risks and challenges associated with e-government that must be addressed by nations to ensure resilient systems and states.

2.2.2.1 Preconditions for E-Government

The UN 2018 E-Government Survey focuses on utilizing e-government to achieve the Agenda 2030 SDGs, with a specific focus on creating resilient and sustainable societies. Accordingly, they identify the following preconditions to ensure successful e-government applications[26]:

1. Political commitment and support of e-government is a critical precondition, and national strategies, policies and plans specifically relating to ICT and e-government need to be established and incorporated at a national level. Furthermore, the national strategies and plans must be aligned with the UN 2030 Agenda and must define clear monitoring and review structures.
2. Technical standards and guidelines defining data sharing, staff requirements, and organizational capacity and responsibility are required to maximize the potential of e-government by ensuring interoperability and digital transactions across the public sector to streamline public administration.
3. Public trust and buy-in are another critical precondition to e-government. This can be increased through clearly defined strategies and frameworks that define a blueprint for public services in terms of performance expectations, basic competence and efficacy, and inclusivity, thereby increasing the transparency and accountability of e-government activities.
4. ICT infrastructure is an essential component in ensuring access to online public services to all population groups and should enable tailor-made service delivery to targeted groups. Obviously, Internet connectivity is a prerequisite to enable citizens to access online public services and as such is a cornerstone in implementing any e-government applications.

---

[23]UN Agenda 2030 SDG 16: "Promote peaceful and inclusive societies for sustainable development, provide access to justice for all and build effective, accountable and inclusive institutions at all levels".

[24]UN Agenda 2030 SDG 17: "Strengthen the means of implementation and revitalize the global partnership for sustainable development".

[25]United Nations Department of Economic and Social Affairs, *E-Government Survey 2018— Gearing E-Government to Support Transformation towards Sustainable and Resilient Societies*, p. 1.

[26]Ibid.

5. Ensuring inclusive public services and administration is more challenging in developing countries that have rural populations and poor ICT infrastructure. In these contexts, non-digital service delivery channels (including post offices, call centres, television, radio and face-to-face services) still form a crucial tool of governance and new strategies must be cognizant of their importance. New e-government strategies can aim to develop a mix of both digital and non-digital service channels but should leverage ICT to support front-line staff (a successful example of such a strategy is Portugal's implementation of Citizen Shops as discussed in Sect.2.4.1.4).

6. Governments must provide a supportive environment to facilitate partnerships relating to e-government at the global, regional and national levels, including both private actors (through PPPs) and multi-stakeholder partnerships.

7. Governments need to abandon outdated traditional service models and focus on user-centricity and co-creation to realize new online public service delivery mechanisms. As a result of ICT, the capacities of civil society and the private sector to assist governments in addressing socio-economic challenges have increased substantially and these actors must be encouraged and promoted to fully maximize their impact.

2.2.2.2 Risks and Challenges Facing E-Government

The 2018 Survey report notes that "environmental stresses and disasters, socio-economic and government risks, as well as those related to technologies themselves"[27] can threaten the role of e-government in promoting the SDGs.[28] With economic, social or political turbulences, public services may become strained due to a diversion of resources away from public administration procedures. As a result, public services can break down entirely thereby negatively impacting progress towards achieving the SDGs.

If societies do not have strong resilience, this can lead to weakened state capacity and may put a nation's population at risk. As such, governments must be cognizant of the threats posed by natural and human-made disasters to e-government systems. As such, e-government systems and the ICT infrastructure supporting them need to be resilient and supplemented with adequate response strategies.

Other risks facing e-government arise from the rapid pace at which technology advances and the myriad ways in which it can be misused or distorted. The formulation of national policies and legal frameworks is a lengthy process, which is often too slow to adequately react to new technological breakthroughs in a timely manner. Other issues relating to the technology that enables e-government include[29]:

1. Concerns relating to data protection and privacy,

---

[27] Ibid., p. 20.

[28] Ibid.

[29] Ibid.

2. Susceptibility to cybersecurity threats and the potentially catastrophic repercussions of hacking of critical infrastructures,
3. Propagation of false news, hate speech and terrorist propaganda through online social media channels,
4. The influence of targeted social media campaigns in the democratic process by exploiting people's base fears and controlling the online advertisements to which they are exposed,
5. In rural areas, with poor ICT infrastructure, e-government may become exclusive to certain population groups unless the digital divide is addressed and overcome.

2.2.3 Ensuring Inclusive E-Governance

Research has confirmed that varying access to technology contributes to socio-economic stratification and, as such, ensuring an inclusive society requires overcoming digital divides.[30] E-government needs to ensure that access to online public services and platforms is made available to the entirety of the population and policy-makers must prioritize inclusion in terms of access and usage.

It is also important to note that digital divides in today's world are no longer only arising from poor ICT infrastructures, lack of access to hardware or connectivity issues. While basic access or connectivity remains the largest digital divide in rural Africa, there are numerous different digital divides,[31] which include: affordability, age, bandwidth, content, disabilities, education/literacy issues, migration, location, mobile access, and Internet connection speeds.[32] Many of these new digital divides are as a result of local contextualized issues and require specific remedial strategies. Furthermore, the type of divides facing nations varies depending on the progress of a nation's digital development. In low-income countries, basic infrastructure upgrades are first required to ensure access to the Internet, before other divides such as bandwidth or speed issues can be addressed.[33]

Affordable, high-speed, high-bandwidth Internet access is a basic requirement to fully realize the potential of e-government. This remains a large challenge in developing countries, where a lack of ICT infrastructure results in large portions of the population having no access at all to the Internet, and even once basic connections are provided, the low speeds and high cost can make going online prohibitive. For

---

[30] M. Warschauer, *Technology and Social Inclusion: Rethinking the Digital Divide*, MIT Press, 2004.

[31] See, for example: C. Yoon, "Digital Africa: An Analysis of Digital Trends in Africa and Their Driving Factors", in *Space Fostering African Societies: Developing the African Continent through Space, Part 1*, ed. A. Froehlich (Cham: Springer, 2020), 109–134; and C. Kotze, "Addressing the Un-Addressed: Opportunities for Rural-Africa", in *Space Fostering African Societies: Developing the African Continent through Space, Part 1*, ed. A. Froehlich (Cham: Springer, 2020), 153–174.

[32] K. Andreasson, *Digital divides: the new challenges and opportunities of e-inclusion*, CRC Press, 2015.

[33] United Nations Department of Economic and Social Affairs, *E-Government Survey 2018—Gearing E-Government to Support Transformation towards Sustainable and Resilient Societies*, 2018.

example, in 2016 it was recorded that in low-income countries only twelve out of every 100 people had access to the Internet.[34]

Mobile devices are an opportunity to increase the number of users in low-income countries as they require less fixed-line infrastructure and have been identified as enablers for developing countries to leapfrog into mobile-only solutions. However, as shown in Sect. 2.2.1, the cost of mobile Internet connectivity can be prohibitively high within Africa. Furthermore, new mobile technologies, such as 5G mobile networks still require the installation of fibre optic networks, which are fixed-line technologies and, as such, developing countries must continue with their commitments to improving their ICT infrastructures to allow for inclusive e-government. Well-defined national broadband and digital strategies should be adopted to facilitate bridging the connectivity divide. In this regard, research has shown that countries that have adopted such strategies exhibit higher penetration rates.[35]

Mobile connection speeds facilitated by 3G networks are deemed the "minimum speed required for smart data functions"[36] and therefore are a requirement for e-government. While the 2018 UN Survey report notes that the population covered by a 3G network was 85% globally in 2018, it should be noted that within Africa this value is considerably lower.[37] The 2019 GSMA Intelligence report on The Mobile Economy in Sub-Saharan Africa (SSA) noted that in 2016, 70% of all mobile connections in SSA were through 2G networks, and only 28% were through 3G networks. This number increased in 2019, to approximately 45%. The GSMA Intelligence report reveals other statistics that highlight just how limited mobile Internet connectivity is within Africa. In 2018, in the SSA region, there were 774 million SIM card connections (representing a 74% penetration rate), however in the same year there were only 239 million mobile Internet users (representing a 23% penetration rate).[38] These statistics reveal that in the SSA region, less than one quarter of the population had access to mobile Internet in 2018 and of those citizens who did have mobile Internet access, not all of them had access to 3G network speeds, which is the minimum requirement for smart data applications. The majority of academic studies present a pessimistic view towards digitization in Africa and overcoming the associated digital divides. However, recent work has revealed a positive trend in digitization in Africa, highlighting an increase in the number of Internet users and the number of mobile subscriptions in African countries exhibiting positive economic and social development levels (reflected by higher economic growth rates, lower inflation rates

---

[34]World Bank, DataBank (online portal), https://databank.worldbank.org/home.aspx, accessed: 8 December 2019.

[35]United Nations Department of Economic and Social Affairs, *E-Government Survey 2018— Gearing E-Government to Support Transformation towards Sustainable and Resilient Societies*, 2018.

[36]Ibid.

[37]Ibid.

[38]GSM Association, *The Mobile Economy in Sub Saharan Africa*, United Kingdom, 2019.

and increased literacy and access to electricity levels). Access to electricity has been shown to be one of the major prohibitors to Internet access in Africa.[39]

New technologies and service providers have emerged to address the lack of Internet connectivity across Africa—for example Facebook is investigating using solar powered, uncrewed aerial vehicles to provide Internet coverage to the whole of the continent. Similarly, numerous space actors are developing large constellations of Low Earth Orbit (LEO) satellites to provide Internet coverage across the entire globe. These new technologies are discussed in detail in Chap. 3.

Another important digital divide in developing countries is the gender gap. The ITU and GSMA organization conducted research in 2010 that determined that in developing countries, women are 21% less likely to own a mobile phone than men.[40] The GSMA *Mobile Gender Gap Report* from 2019 reveals that women's mobile phone ownership in low- and middle-income countries has since increased with women being 10% less likely than men to own a mobile device in these countries (and 23% less likely to use mobile Internet than men).[41] However, the mobile gender gap varies greatly by region—the worst region was South Asia, where "women are 28% less likely than men to own a mobile and 58% less likely to use mobile Internet".[42] Other key findings from the report include[43]:

- Mobile Internet represents the primary means of connection for low- and middle-income countries, especially for women.
- The key barriers to mobile ownership and Internet use for women include: literacy, affordability, digital skills and security concerns.
- 80% of women currently own a mobile phone in low- and middle-income countries, representing an increase in 250 million women since 2014, and 48% of women in these regions use mobile Internet.

Another contributing factor to the online gender gap is a lack of online content targeted towards female audiences.[44]

Additional contributors to the digital divide are a lack of awareness or perceived benefit and a lack of online content in local languages. As such, governments need to raise the awareness of their online presence, highlight its associated benefit, and

---

[39]Christopher Yoon, *Digital Africa: An analysis of digital trends in Africa and their driving factors*, Space Fostering African Societies—Developing the African Continent through Space Part 1, Springer, Switzerland, 2020, ISBN 978-3-030-32929-7, https://doi.org/10.1007/987-3-030-329 30-3.

[40]GSM Association, *Women & Mobile: A study on the mobile and phone gender gap in low and middle-income countries*, United Kingdom, 2010.

[41]GSM Association, *Connected Women—The Mobile Gender Gap Report 2019*, United Kingdom, 2019.

[42]Ibid.

[43]Ibid.

[44]United Nations Department of Economic and Social Affairs, *E-Government Survey 2018—Gearing E-Government to Support Transformation towards Sustainable and Resilient Societies*, 2018.

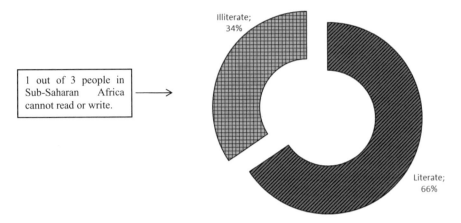

Fig. 2.12 Illiteracy rate in Sub-Saharan African for people aged 15 and older

ensure online content is available in multiple languages to ensure inclusivity in e-government.[45] Additional barriers for people with disabilities exist—for example people with impaired vision can only access websites that are compatible with screen readers or similar technologies. In the 2018 Survey, less than 40% of the national websites for member states were fully accessible to people with disabilities.[46]

Digital literacy constitutes another digital divide preventing fully inclusive e-government as vast numbers of people across the world lack the digital skills to actually connect to the Internet and access online public services. This is generally most prevalent in older demographics who use ICTs less regularly. This can be addressed by establishing face-to-face kiosks with trained civil servants who assist such people in obtaining online services. Such an approach was implemented in Portugal using the Citizen Spot initiative, launched in 2014.[47]

In Africa and other developing countries, it is not only digital literacy, but literacy in general, that forms a barrier for ensuring inclusivity in e-government. The World Bank and UNESCO data show that in SSA in 2018, there was a literacy rate of 65.6% for people aged 15 and older,[48] which means that one in three people in the region cannot read and write (as highlighted in Fig. 2.12).

Governments need to be aware of these digital divides and identify specific remedial plans to address them in their national digital strategies. Furthermore, when implementing e-government activities and services, thought must be given to people

---

[45] Ibid.

[46] Ibid.

[47] Ibid.

[48] World Bank, DataBank, *Literacy Rate—Adult Total (% of People ages 15 and above)*, https://data.worldbank.org/indicator/SE.ADT.LITR.ZS?end=2018&locations=ZG-1W-Z4-8S-Z7-ZJ&start=2018&view=bar, accessed: 8 December 2019.

who do not have access to the Internet to ensure that the "digital first" approach does not unintentionally exclude a nation's citizens from participation.

2.2.4   Cybersecurity Issues Associated with E-Government

The ITU noted in its 2018 *Global Cybersecurity Index (GCI)* report that by the end of 2018, there were more than 3.9 billion people with access to the Internet, representing a 51% penetration rate, which is expected to increase to 70% by 2023. Furthermore, due to the increase in ICT utilization in government systems, the ITU estimated that the cost of cybercrimes by the end of 2019 will reach US$ 2 trillion.[49] With governments offering more public services online, the threats of online disruptions arising from cyberattacks or natural disasters are ever increasing. The interdependencies of online public platforms means that such disruptions can directly impact governments' ability to provide services in the health, safety, security and social sectors and may negatively impact the proper functioning of a country's economy.

Cyberattacks in general across the globe are on the increase, with far reaching results—for example the *WannaCry* ransomware attack in 2017 affected more than 150 countries. The National Health Service (NHS) in the United Kingdom was greatly impacted—more than one third of NHS organizations were directly affected, key medical equipment was damaged or destroyed, putting patient safety directly at risk. Furthermore, cyberattacks can result in huge economic losses and weaken a nation's gross domestic product (GDP), as with the Netherlands in 2017 where e-crime and identity theft resulted in a loss of €10 billion.[50] These numbers highlight the necessity of effective cybersecurity to ensure resilient and secure e-government systems.

Cybersecurity is also paramount to ensure that citizens' private data and sensitive information are protected from potential cyberattacks while at the same time ensuring data privacy rights are not impeded. This requires coherent, comprehensive and responsive cybersecurity legislation and policies which keep abreast with advances in the ICT sector. The ITU states that "e-governance can only function safely if cybersecurity is implemented effectively" and that cybersecurity also has an impact on "the overall developments in ICT" infrastructure.[51]

The ITU's GCI can be employed to assess Africa's progress and commitments towards addressing cybersecurity risks. The GCI was included under the ITU Plenipotentiary Resolution 130 to assist in strengthening the role of the ITU in increasing confidence and security in the use of ICT as per the Global Cybersecurity Agenda

---

[49]ITU, *Global Cybersecurity Index (GCI) 2018*, ITU Publications, Switzerland, 2019, ISBN 978-92-28201-1 (electronic version).

[50]United Nations Department of Economic and Social Affairs, *E-Government Survey 2018—Gearing E-Government to Support Transformation towards Sustainable and Resilient Societies*, 2018.

[51]ITU *Global Cybersecurity (GCI) Index 2018*, 2019.

(GCA) framework launched in 2007. The GCI is a composite index based on 25 indicators across five separate pillars identified in the GCA, with the aim of measuring: "the type, level and evolution over time of cybersecurity commitment in countries and relative to other countries; progress in cybersecurity commitment of all countries from a global perspective; progress in cybersecurity commitment from a regional perspective; the cybersecurity commitment divide".[52]

The five pillars and their relevant indicators are listed below:

1. **Legal**: Evaluates the legal institutions and frameworks specifically relating to cybersecurity and cybercrimes. Indicators include[53]:

   (a) Cybercrime legislation
   (b) Cybersecurity regulation
   (c) Containment/curbing of spam legislation

2. **Technical**: Assesses the level of technical institutions and frameworks directly related to cybersecurity. Indicators for this pillar include[54]:

   (a) Computer Emergency Response Team (CERT) and/or Computer Security Incident Response Team (CSIRT) organizations
   (b) Standards implementation frameworks
   (c) Standardization body or institution
   (d) Technical mechanisms and capabilities deployed to address spam
   (e) Use of cloud storage for cybersecurity purposes
   (f) Child Online Protection (COP) mechanisms.

3. **Organizational**: Measures mechanisms for implementing and coordinating policies and strategies for cybersecurity development at a national level. The indicators used to evaluate the organizational pillar are[55]:

   (a) National Cybersecurity Strategy
   (b) Responsible Agency
   (c) Cybersecurity Metrics.

4. **Capacity Building**: Assesses agencies focused on fostering capacity through research and development, education and training initiatives, ensuring certification for professionals and other public agencies. The relevant indicators are[56]:

   (a) Public awareness campaigns
   (b) Frameworks for certification and accreditation of cybersecurity professionals

---

[52] Ibid.
[53] Ibid.
[54] Ibid.
[55] Ibid.
[56] Ibid.

(c) Professional training courses in cybersecurity
(d) Educational programmes or academic curricula in cybersecurity
(e) Cybersecurity research and development programmes
(f) Incentive mechanisms.

5. **Cooperation**: Investigates existing partnerships, cooperative frameworks and information sharing networks within a country. The indicators for the cooperation pillar are[57]:

(a) Bilateral agreements
(b) Multilateral agreements
(c) Participation in international associations
(d) PPPs
(e) Inter-agency/intra-agency partnerships
(f) Best practices.

In this way, governments can use the GCI not only as a benchmark, but also as a guide to identify shortfalls within their cybersecurity environment and identify and address the associated threats. The GCI scores of member states are used to classify them based on their commitment to cybersecurity: Countries in the "High" group demonstrate a high commitment across all five pillars, the "Medium" group "have developed complex commitments and engage in cybersecurity programmes and initiatives", while the "Low" group have only just begun to commence their commitments in relation to cybersecurity. Figure 2.13 demonstrates clearly the low commitment levels towards cybersecurity throughout Africa. However, four African countries did achieve "High" classifications, namely Egypt, Kenya, Mauritius and Rwanda.

Table 2.4 shows the GCI scores for the top ten countries in Africa (the highest possible score is 1.00), as well as their rankings out of all 193 UN member states.

Mauritius performed well, ranking 14th globally and scoring the highest regional score for the organizational pillar. Mauritius has established numerous initiatives that focus on cybersecurity including: the National Cybersecurity Strategy, the National Cybercrime Strategy, the National Cyber Incident Response Plan and the Computer Emergency Response Team of Mauritius (CERT-MU). Furthermore, to ensure resilience, Mauritius has established the National Disaster Cybersecurity and Cybercrime Committee (comprised of both public and private actors) which facilitates and supports governance decisions during a cyber crisis within the country.[58]

The performance of Africa as a region in each of the five pillars in the GCI report is addressed below:

1. **Legal**: Numerous African countries have established cybercrime and cybersecurity legislation, and in general the performance of African countries in the legal

---

[57] Ibid.
[58] ITU *Global Cybersecurity Index 2018*, 2019.

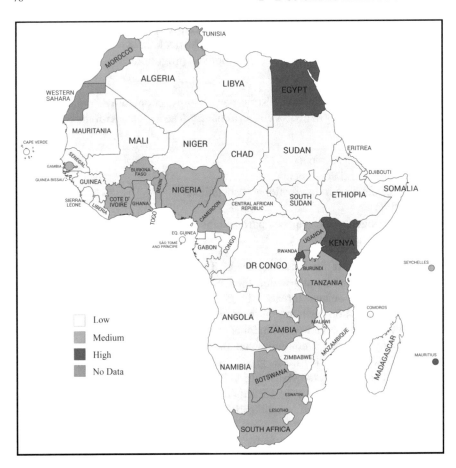

Fig. 2.13 Global GCI classification of member states' commitment to cybersecurity (Ibid., p. 13)

Table 2.4 Top ten African countries in the 2018 GCI Report

2018 Rank in Africa	Country	GCI score	Classification	2018 Global rank
1	Mauritius	0.880	High	14
2	Egypt	0.842	High	23
3	Kenya	0.748	High	44
4	Rwanda	0.697	High	49
5	South Africa	0.652	Medium	56
6	Nigeria	0.650	Medium	57
7	Tanzania	0.642	Medium	59
8	Uganda	0.621	Medium	65
9	Tunisia	0.536	Medium	76
10	Benin	0.485	Medium	80

Based on information contained within the ITU *Global Cybersecurity Index Report* 2018

framework pillar is adequate. The number of African countries that have shown commitment to the indicators for the legal pillar are[59]:

(a) 38 countries have cybercrime legislation.
(b) 37 countries have regulations relating to cybersecurity.
(c) 22 countries have legislation relating to curbing spam.

2. **Technical**: African states fare poorly in the indicators relating to the technical pillar, with the exception that as a region they have the second highest number of states that have implemented COP measures (Europe is first with 38 countries having implemented such measures). The technical pillar is a key area where African countries need to increase their efforts. The number of African countries that have demonstrated commitment to the indicators for the technical pillar are[60]:

(a) Only 13 African countries have national CERTs.[61]
(b) Only nine countries have technical national cybersecurity standards, which should be improved especially considering there are 18 African countries with national standardization bodies/agencies.
(c) Only five African countries exhibit the technical mechanisms and capabilities to address spam.
(d) Only ten African countries exhibit commitment to using the cloud to increase the resilience of their online systems, which is of concern.
(e) 19 African countries have implemented COP measures, which is the second highest number of all regions.

3. **Organizational**: Broadly speaking, all regions exhibited a significant number of countries that have implemented a national agency for cybersecurity. However, Africa still needs to work to increase its performance in this regard. There are 22 countries with a national agency for cybersecurity within Africa, but only 14 countries have a national cybersecurity strategy or framework,[62] and only ten have identified and implemented metrics for measuring national cybersecurity measures.[63]

4. **Capacity Building**: Globally there was a positive increase in capacity building initiatives, specifically relating to increasing the local or homegrown cybersecurity industry within states. However, this is another area where African countries

---

[59]Ibid.

[60]Ibid.

[61]As of March 2019, the ITU noted only 13 African countries have national Computer Emergency Response Teams, specifically: Burkina Faso, Cameroon, Ethiopia, Ghana, Ivory Coast, Kenya, Mauritius, Nigeria, Rwanda, South Africa, Tanzania, Uganda and Zambia.

[62]The ITU lists the following 14 African countries who have a national cybersecurity strategy or framework: Burkina Faso, Gambia, Ghana (Draft), Kenya, Malawi, Mauritius, Nigeria, Rwanda, Senegal, Sierra Leone, South Africa, Tanzania, Uganda, and Zambia (Draft). Botswana is currently developing their national cybersecurity framework.

[63]ITU *Global Cybersecurity Index 2018*, 2019.

need to improve. The indicators for this pillar along with the number of African states that have established commitments for each are shown below[64]:

(a) Cybersecurity public awareness campaigns: 25 countries,
(b) Frameworks for certification & accreditation: 9 countries,
(c) Professional courses in cybersecurity: 17 countries,
(d) Educational programmes & academic curricula: 22 countries,
(e) Research & development initiatives: 22 countries,
(f) Incentive mechanisms for promoting capacity: 6 countries,
(g) Homegrown cybersecurity industry: 11 countries.

5. **Cooperation**: Africa's participation in international associations relating to cybersecurity is fairly strong (with 32 countries), however at the end of 2018 there were only twelve countries with bilateral agreements, 17 with multilateral/international agreements and only 14 countries with inter-agency agreements. Furthermore, PPPs relating to cybersecurity initiatives are only present in 16 African countries. Africa needs to focus on increasing their participation in cooperative approaches to cybersecurity, as ensuring secure and safe e-government systems requires input from various sectors and disciplines and requires cooperation between governments, agencies, private sector actors, civil society and academia.[65]

This summary reveals that while Africa is starting to demonstrate increased awareness and commitment to cybersecurity, more work is required to ensure resilient e-government systems on the continent. The starting point is for all African nations to draft and adopt legislation and policies that focus specifically on cybersecurity and cybercrimes. Countries should aim to harmonize such legislation on a regional, continental and international level to allow for interoperable responses and facilitate the international fight against cybercrimes.

Once African states have implemented the necessary policies, frameworks and legislation, they should focus on establishing national dedicated agencies or organizations that are responsible for defining national cybersecurity development plans and response strategies. Such frameworks should identify critical information infrastructures (CIIs) and impose specific cybersecurity measures relating to them to ensure resilience. The 2018 GCI findings highlight that Africa is especially weak when it comes to technical capacity relating to cybersecurity which must be addressed as a matter of urgency. African states must adopt minimum security criteria for e-government systems and establish an accreditation methodology to ensure safe and secure online systems. Countries should establish CERT/CSIRT departments that report to their national responsible agency and are responsible for identifying, managing and responding to cyber threats as well as continuously reviewing the security of national online platforms. This is only possible if significant efforts are made to increase the capacity and human capital within the sector. This necessitates

---

[64]Ibid.
[65]ITU *Global Cybersecurity Index 2018*, 2019.

large scale awareness campaigns coupled with optimized training and education schemes, nationally recognized accreditation programmes and incentives to attract more professionals to the sector.

Lastly, cooperation and collaboration are critical aspects of managing and responding to cyber threats and all cybersecurity programmes must promote cooperation on a regional, organizational, national and international level. Including actors from the private sector as well as experts from academia is vital to ensuring the development of successful cybersecurity schemes.

2.3   Emerging Technologies in the E-Government Sector

Numerous new technologies afforded by the fourth industrial revolution are being investigated for their applications within e-government and e-democracy. These include the Internet of things (IoT), big data analytics, artificial intelligence (AI), machine learning and blockchain in various combinations.[66,67] These technologies allow for new e-government models, called digital government, which is defined as "the creation of new public services and service delivery models that leverage digital technologies and governmental and citizen information assets. The new paradigm focuses on the provision of user-centric, agile and innovative public services."[68]

2.3.1   Blockchain

Blockchain has been lauded as a potentially disruptive technology for e-government applications and is currently one of the most used distributed ledger technologies (DLTs). According to the European Commission's Joint Research Centre (JRC), "distributed ledger technology refers to the protocols and supporting infrastructure that allow computers in different locations to propose and validate transactions and update records in a synchronized way across a network".[69]

Although DLTs have been around since 2004, blockchain gained international attention through the independent digital currency, Bitcoin, in 2008, which used DLT to create a purely peer-to-peer electronic transaction network, thereby nullifying the

---

[66]S. Terzi et al., *Blockchain 3.0 smart contracts in e-government 3.0 applications*, CERTH/ITI and Aristotle University of Thessaloniki, Greece, 2019.

[67]S. Myeong and Y. Jung, *Administrative reforms in the fourth industrial revolution: the case of blockchain use*, Sustainability Journal, https://doi.org/10.3390/su11143971, July 2019, www.mdpi.com/2071-1050/11/14/3971, accessed: 28 February 2020.

[68]D. Alessie et al., *Blockchain for digital government*, Report EUR 29677 EN, Publications Office of the European Union, Luxembourg, 2019, ISBN 978-92-76-00581-0, https://doi.org/10.2760/942739, JRC115049.

[69]Ibid.

need for a central authority body.[70] The blockchain ledger stores a public record of every transaction performed by Bitcoin users, linked using cryptography. The records are stored as "blocks" which contain a timestamp, transaction data and a cryptographic hash of the previous block.[71] In this way, the ledger cannot be tampered with or edited by any user. Transactions are processed and verified by peers on the network in a de-centralized manner, thereby drastically improving transaction efficiency and security. Since the data are not stored on a single central server and are distributed across all the users, cyberattacks on a single user have a negligible impact on the overall blockchain.[72]

Using blockchain in this manner (i.e. for financial transaction automation without a central authorizing body) is referred to as Blockchain 1.0 whereas Blockchain 2.0 extends the application of blockchain beyond the financial sector through the use of smart contracts. Smart contracts essentially constitute a self-executing computer program that runs on blockchain and allows participating parties to agree on specific conditions of the contract and actions to occur once it is completed. Smart contracts are executed automatically with no human input and are irrevocable and binding once performed. Smart contracts hence allow for regulations and policies to be translated into code, which is executed automatically once specific conditions are met, thereby opening up the application of blockchain technology to healthcare, education, government services, welfare grants, charities, real estate management, insurance and banking.[73,74]

The core benefits of blockchain technology (and DLT in general) are listed below[75,76,77]:

1. The decentralized, shared nature of the technologies results in zero down-time and allows for ledgers and records to be fully disclosable and trackable (even on large scales).[78]
2. The implementation of smart contracts establishes trust between parties and ensures immutable non-repudiation of transactions.
3. The elimination of a central server ensures the system is robust against cyber-attacks as there is no hackable single point of failure within the network.

---

[70]Ibid.

[71]S. Myeong and Y. Jung, *Administrative reforms in the fourth industrial revolution: the case of blockchain use*, 2019.

[72]Ibid.

[73]S. Terzi et al., *Blockchain 3.0 smart contracts in e-government 3.0 applications*, 2019.

[74]D. Alessie et al., *Blockchain for digital government*, 2019.

[75]Ibid.

[76]S. Terzi et al., *Blockchain 3.0 smart contracts in e-government 3.0 applications*, 2019.

[77]R. Qi et al., *Blockchain-powered internet of things, e-governance and e-democracy*, 2017. In: T. Vinod Kumar (editor), *E-Democracy for Smart Cities—Advances in 21st Century Human Settlements*, Springer, 2017, https://doi.org/10.1007/978-981-10-4035-1_17.

[78]D. Alessie et al., *Blockchain for digital government*, 2019.

4. The integrity of data stored using blockchain is guaranteed as previous entries on the ledgers cannot be edited or changed (new records are appended to the list and are cryptographically linked to the previous record).
5. Verification and processing of transactions is performed by the peer-to-peer network, which firstly increases the efficiency and scalability of the system, and secondly, removes the need for a central authorization party, thereby ensuring a truthful and honest ledger that cannot be corrupted.

The decentralized nature and inherent security of blockchain systems reduces the usual associated costs of establishing large networks with central servers. As such, applying this technology in government applications reduces the cost, time and complexity of establishing inter-departmental information exchanges. Furthermore, public distributed ledgers reduce corruption and increase the transparency and accountability of government registries, which can lead to an increase in trust from citizens. Using smart contracts can also assist with automatically enforcing regulations and compliance using online systems thereby eliminating the numerous person-hours and associated costs of employing public servants and specialists to perform such tasks.[79,80]

Governments from countries all across the globe are researching and developing novel ways of applying blockchain to assist in public services and administration. A few examples from the USA are listed below[81]:

- The Defense Advanced Research Projects Agency (DARPA) is creating a secure messaging system for facilitating communications between intelligence officers and the Pentagon using a blockchain-based platform.
- The Food and Drug Administration (FDA) is performing a pilot study investigating using blockchain to track and verify prescription medications.
- The United States Department of Health and Human Services (HHS) is researching using blockchain and AI to reduce operational costs and process times and has invested US$ 49 million in the project.
- The Naval Air Systems' Fleet Readiness Center is developing a blockchain system to track and assess aviation parts through the various phases of their life cycles.

Numerous countries within Africa are also investigating blockchain and DLT, with many choosing to focus on using the technologies to assist with land and property registries (including Ghana, Rwanda, South Africa and Zambia). Tanzania used blockchain technology to audit their public payroll systems, exposing over 10,000 ghost workers with an associated cost of approximately US$ 200 million per month.[82]

---

[79]Ibid.

[80]R. Qi et al., *Blockchain-powered internet of things, e-governance and e-democracy*, 2017.

[81]ConsenSys, *Which governments are using blockchain right now?*, Enterprise Blockchain, November 2019, https://consensys.net/blog/enterprise-blockchain/which-governments-are-using-blockchain-right-now/, accessed: 28 February 2020.

[82]ConsenSys, *Which governments are using blockchain right now?*, Enterprise Blockchain, November 2019, https://consensys.net/blog/enterprise-blockchain/which-governments-are-using-blockchain-right-now/, accessed: 28 February 2020.

The European Commission has been funding blockchain projects through the Financial Program 7 and Horizon 2020 research projects and provides up to € 340 million in funds, with a specific focus on identification management and processing, taxation reporting, e-voting, regulatory compliance and managing development aid funds.[83]

The city of Zug in Switzerland is home to "Crypto Valley"—the largest blockchain ecosystem in the world. The city, along with the Institute for Financial Services Zug (IFZ) of Lucerne University, Luxoft and the uPort decentralized identity platform trialled using blockchain (specifically Ethereum) and DLT for enabling online voting in a direct democracy. The specific issue in the pilot study revolved around allowing fireworks at a local festival, however the project proved to be successful in that voter identification and validation was achieved without the need for any intermediaries or external vote-counting institution. However, it is important to note the scale of the pilot project only involved 350 registered users.[84]

2.3.2 E-Voting

Blockchain is a technology well-suited to overcoming many of the problems with e-voting due to its decentralized nature and secure, immutable distributed ledgers, and satisfies all the requirements for e-voting systems. E-voting is not a new concept and has received much attention due to its ability to increase the security, verifiability, transparency and efficiency of the voting process. The typical requirements for e-voting systems include[85]:

1. Ensuring the privacy of the voter and concealing their vote,
2. Ensuring that only registered voters can vote,
3. Ensuring each registered voter votes only once,
4. Achieving verifiability to confirm if a specific vote was counted and counted correctly, while simultaneously ensuring that a voter cannot demonstrate to a third party the way they voted, in order to prevent voter coercion,
5. Ensuring the voting process is easy to perform and available to all eligible voters,
6. Promoting trust in the vote tallying process.

The first "electronic voting" systems comprised computer-aided counting systems using punched-card ballots in the 1960s. These computer counting systems, also referred to as paper-based electronic voting systems, were improved to connect to central servers at polling stations to scan and tally votes. Direct Recording Electronic

---

[83]D. Alessie et al., *Blockchain for digital government*, 2019.

[84]ConsenSys, *Zug digital ID: Blockchain case study for government issued identity*, Enterprise Ethereum, https://consensys.net/blockchain-use-cases/government-and-the-public-sector/zug/, accessed: 28 February 2020.

[85]L. Rura et al., *Implementation and evaluation of steganography and visual cryptography*, Journal of Computing and Information Technology, 2017, https://doi.org/10.20532/cit.2017.1003224.

(DRE) voting systems were the next development in the sector and saw electronic voting machines (EVMs) being incorporated inside voting booths, allowing voters to cast their vote using either touchscreens or electro-mechanical buttons connected to a display screen.[86] DRE systems have certain advantages—for example ballots do not have to be physically printed (resulting in cost savings), offering the voting interface in numerous languages, accommodating people with disabilities, reducing the time required for counting ballots, lessening the number of spoilt ballots, and reducing the risk of double voting.

DRE systems have, however, received substantial criticism with critics claiming that EVMs are susceptible to fraud and tampering and generally exhibit a lack of independent auditability.[87] In 2006 Ireland stopped its development of EVMs due to security concerns.[88] Similarly, in 2007 the Netherlands ceased using EVMs after the security weaknesses in the machines were highlighted by a group of hackers. In 2008, the German Federal Constitutional Court deemed the use of voting machines for federal elections unconstitutional; stating that the correctness of the results obtained using such machines is not publicly verifiable.[89]

Extensive research has been performed to facilitate online voting, allowing citizens to cast their ballots over the Internet with the aim of increasing voter participation. A key element for online voting systems is ensuring an accurate and reliable voter identification process (which requires an electronic identity infrastructure) as well as providing publicly verifiable results.[90] Norway trialled an online voting platform between 2011 and 2013, however the project was terminated due to poor public perception of the security of the online vote.[91] A report by Canada's cybersecurity agency, the Communications Security Establishment (CSE), noted that "in 2018, half of all advanced democracies holding national elections had their democratic process targeted by cyber threat activity. This represents about a three-fold increase since 2015".[92]

---

[86]K. Khan et al., *Secure digital voting system based on blockchain technology*, NED University of Engineering Technology (Pakistan) and the University of West London (United Kingdom), 2018, https://core.ac.uk/download/pdf/155779036.pdf, accessed 3 March 2020.

[87]National Institute of Standards and Technology, *Security and Subcommittee of the Technical Guidelines Development Committee—Recommendations for Software Independence in Voluntary Voting Systems Guidelines*, US Department of Commerce, United States of America, 2007.

[88]L. Omarjee, *E-voting: Which countries use it, where has it failed and why?* Fin24, May 2019, www.fin24.com/Economy/e-voting-which-countries-use-it-where-has-it-failed-and-why-20190510, accessed: 3 March 2020.

[89]C. Henrich, *Improving and analysing Bingo Voting (Doctorate Thesis)*, Karlsruhe Institute of Technology, Germany, 2012.

[90]Microsoft Corporate Blogs, *Electronic voting: What Europe can learn from Estonia*, Microsoft Corporate Blogs—EU Policy Blog, Microsoft, May 2019, https://blogs.microsoft.com/eupolicy/2019/05/10/electronic-voting-estonia/, accessed: 3 March 2020.

[91]L. Omarjee, *E-voting: Which countries use it, where has it failed and why?* Fin24, May 2019.

[92]Communications Security Establishment, *2019 Update: Cyber threats to Canada's democratic process*, Government of Canada, 2019.

Conversely, in Estonia, 31.3% of citizens voted online in the 2014 European Parliament Elections using the i-Voting system.[93] It became the first country in the world to use online voting systems in its 2005 national election and has been using them since, saving over 11,000 working days per election as a result. The Chairman of the Estonian Electronic Voting Committee stated that in the last election in March 2019, 44% of voters used the online system and noted that although establishing an online voting system does require a substantial capital investment, once it is set up it is significantly cheaper than paper-based systems. He also noted that the majority of people voting online are aged between 25 and 45 and indicated a correlation between online voters and the distance to their nearest polling station.[94]

Other examples of e-voting platforms that address the issues of voter identification, vote verification, voter anonymity and end-to-end auditability include: Bingo Voting, Helios Voting, the DRE-i voting system, DRE-ip voting system, and the STAR-Vote voting system.[95]

The above Estonian example highlights how advantageous online e-voting systems can be when they are implemented correctly and are able to overcome security, transparency and auditability concerns. However, the majority of existing e-voting systems still require either an independent, trusted authority to perform tallying operations or to distribute the trust among numerous tallying authorities. Helios is an illustration of the latter, using threshold cryptography as a means of incorporating numerous tallying authorities.[96] Blockchain is a new technology that can overcome many of the hurdles facing e-voting systems and simultaneously provide self-tallying, decentralized solutions that are affordable and publicly verifiable.[97]

Examples of using blockchain as a solution for enabling e-voting are listed below:

- "Anonymous Voting by Two-Round Public Discussion" which is based on the two-round Anonymous Veto Protocol (named AV-net), with the introduction of a self-tallying function,[98]

---

[93]The Estonian i-Voting system uses an application to capture a citizen's vote. The identity of the citizen is confirmed using their ID card which is read using a smart card reader or mobile phone. Identity verification is completed using a unique PIN code. Once a vote is completed it is encrypted and sent to the election commission anonymously. Voters can confirm their vote arrived correctly at the election commission server by scanning a QR code through the application, www.valimised. ee/en/internet-voting/stages-i-voting-voter-application.

[94]Microsoft Corporate Blogs, *Electronic voting: What Europe can learn from Estonia*, Microsoft Corporate Blogs—EU Policy Blog, Microsoft, May 2019.

[95]K. Khan et al., *Secure digital voting system based on blockchain technology*, 2018.

[96]P. McCorry et al., *A smart contract for boardroom voting with maximum voter privacy*, Newcastle University, United Kingdom, 2017, https://eprint.iacr.org/2017/110.pdf, accessed: 3 March 2020.

[97]Ibid.

[98]F. Hao et al., *Anonymous voting by two-round public discussion*, The Institute of Engineering and Technology Information Security, 2008, https://doi.org/10.1049/iet-ifs.2008.0127. http://hom epages.cs.ncl.ac.uk/feng.hao/files/OpenVote_IET.pdf, accessed: 3 March 2020.

- "A Smart Contract for Boardroom Voting with Maximum Voter Privacy" was the first successful application of a self-tallying, decentralized protocol named Open Vote Network (OVN) and is based on the Ethereum blockchain,[99]
- "Netvote", which uses the Ethereum network to provide a decentralized voting solution with different applications as the user interface. For example, the application Voter dApp is used by voters for registration and voting and can communicate with biometric readers,[100]
- "Blockchain-Based E-Voting System" is a research project that proposes a novel approach using a private Ethereum blockchain,[101]
- "Agora Technologies" is a company providing an end-to-end verifiable voting solution using a custom blockchain in a multi-layer architecture.[102]

The presidential election in Sierra Leone in 2018 was the first ever national elections that involved blockchain technology, albeit not directly.[103] The Swiss blockchain company, Agora Technologies was accredited and contracted by Sierra Leone's National Electoral Commission (NEC) to act as a trusted, third-party observer to provide an independent count of ballots in the Western District of the country.[104] While this is a limited role for blockchain technology, it does evidence that national governments are seriously considering the new technology and its numerous potentials. In September 2019, Agora Technologies announced a partnership with Onfido[105]—a technology start-up created by three Oxford graduates in 2012. Onfido provides identification verification using AI-based software to determine if a user's government-issued ID card is genuine or fraudulent, compares it to a photograph of the user (captured on a mobile phone) and verifies it against their facial biometrics.[106] The partnership will assist Agora with providing a reliable, fast and easy-to-use voter registration process.

---

[99] P. McCorry et al., *A smart contract for boardroom voting with maximum voter privacy*, 2017.

[100] S. Landers, *Netvote: A decentralized voting platform*, Medium, April 2018, https://medium.com/netvote-project/netvote-a-decentralized-voting-platform-cc2f8227dcaf, accessed: 3 March 2020.

[101] F. Hjálmarssion and G. Hreidarsson, *Blockchain-based e-voting system*, Reykjavik University, Iceland, 2018.

[102] Agora Technologies, *Agora—Technology*, Agora Technologies, Swiss Lab & Foundation for Digital Democracy, www.agora.vote/technology, accessed: 3 March 2020.

[103] M. del Castillo, *Sierra Leone secretly holds first blockchain-audited presidential vote*, CoinDesk, Digital Currency Group, March 2018, www.coindesk.com/sierra-leone-secretly-holds-first-blockchain-powered-presidential-vote, accessed: 3 March 2020.

[104] CoinDesk, *The Sierra Leone vote: What we got wrong*, CoinDesk, Digital Currency Group, March 2018, www.coindesk.com/blockchain-vote-election-sierra-leone-got-wrong, accessed: 3 March 2020.

[105] Finextra Research, *Agora and Onfido partner for blockchain e-voting service*, Finextra Research, September 2019, www.finextra.com/pressarticle/80030/agora-and-onfido-partner-for-blockchain-e-voting-service, accessed: 3 March 2020.

[106] Onfido, *About Us*, https://onfido.com/about/, accessed: 3 March 2020.

2.3.3   Digital Government

Combining blockchain technology with AI and IoT-enabled sensors is a powerful and promising new field with numerous applications in smart cities. An example of the scale of such applications is the Walmart, IBM and Tsinghua University collaboration that seeks to provide quality assurance of food products. Using the integrated blockchain system, individual food products can be tracked from suppliers to grocery stores and then to customers, thereby ensuring the quality and safety of the food product. Blockchain is used to store a myriad of data relating to each individual food item, including the original farm details, batch numbers, processing information, expiry dates, shipping details and storage times and temperatures.[107]

Leveraging the technologies from the fourth industrial revolution allows governments to evolve from centralized e-government systems providing limited services to digital governments that are relationship-focused, providing custom services to individual citizens. Table 2.5 shows how various government objectives and services, enabling technologies, and decision-makers move towards realizing digital government.

The 2019 European Commission report on *Blockchain for Digital Government* noted that while the potential for government transformation using blockchain is great, to date "blockchain has not yet demonstrated to be either transformative or even disruptive innovation for governments".[108] However, it stated that existing projects utilizing the technology have provided clear value for citizens through incremental changes to operational capacities and processing within governments. The report assessed numerous projects throughout Europe and found that the two most common stumbling blocks were (i) incompatibility with existing legacy systems, and (ii) non-compliance with current regulatory and organizational frameworks. As such, new policies need to be developed to address these non-technological barriers and new blockchain projects, policies and frameworks should seek to develop new structures, information systems and standards rather than try to adapt existing systems.

2.4   Specific E-Government Applications and Case Studies

This section details examples of exemplary e-government applications and services both in developed and developing countries.

---

[107] R. Qi et al., *Blockchain-powered Internet of things, e-governance and e-democracy*, 2017.

[108] D. Alessie et al., *Blockchain for digital government*, 2019.

Table 2.5 Progression towards a relationship-focused smart e-government

Category	Centralized e-government	Citizen-oriented e-government	Individual-oriented e-government	Digital government
Objective of E-Gov.	Efficiency of system First-stop shop	Info sharing and connectivity One-stop shop	Open big data, individualized services My Gov.	Relationship-oriented AI platform We Gov.
E-Gov. services	Information provision	Gov. reform and online portals	Platform-based services	Relationship-centric data management
ICT environment	Gov. driven using outsourcing	Gov. driven using outsourcing	Gov. and citizen partnership, deregulation	Autonomous control based on governance
Enabling technologies	Web browser and online storage	Broadband, rich content models	Semantic tech, sensor networks	Blockchain AI
Decision-makers	Political bodies and members of parliament	Gov. professionals and public officials	Individuals, citizens, NGOs	Data-embedded collective intelligence
Role of central gov.	Initiator	Contractor	Mediator	Co-operator
Role of local gov.	ICT and system building	Establishing local gov. portals	Local demand-based personalized services	Platform-based community supported data service delivery
Role of citizen	Information service user	Participation in local government	Active participation and voting	Realization of social value through choice

S. H. Myeong, *A study on the paradigm change of Gov3.0-based e-government service with smart society transition*, Research Report, Korea Information Science Promotion Agency, Korea, 2012)

2.4.1   Exemplary E-Government Applications in the Developed World

Five particular illustrations of innovative and successful e-government applications in developed countries are examined. First, the UK company, IrisGuard, which has developed a biometrically authenticated financial delivery platform currently used to provide both financial services and biometric identification of Syrian refugees is discussed. The second example is the Presidency's Communication Centre in Turkey, which supports citizen participation and inclusion in public services. The third example, Text4Baby in America, is an initiative that issues free SMS messages to users containing relevant healthcare advice for new and expectant mothers in low-income areas. Portugal's Citizen Shops and Citizen Spots is the fourth example, which is a good model for countries attempting to extend the reach of their online services in large geographic areas coupled with poor digital literacy skills. The final example within a developed context is the X-Road open source data exchange layer, which was developed by Estonia, but has been adopted by numerous other countries and provides a safe, reliable and secure method for connecting different information systems and, as such, is a crucial tool for new countries looking to increase their e-governance levels and provide integrated online citizen portals.

2.4.1.1   IrisGuard—Providing Financial Inclusion to Refugees

According to their website, IrisGaurd "is the world's leading supplier of end-to-end iris recognition financial delivery platforms, that since 2001 have been authenticating large-scale humanitarian deployments, delivering on the bold objective to provide financial inclusion and authorization of transactions for refugees and vulnerable populations".[109]

IrisGuard's biometric identification system supports Target 16.9 of SDG 16, which aims to provide a legal identity for all, including free birth registrations.[110] IrisGuard states that approximately 2.5 billion people globally do not have access to financial services and traditional bank accounts. A large contributor to this problem is the lack of a secure identification system. Its biometric identification system uses an iris imager that captures a greyscale image of a person's iris (the ring-shaped coloured membrane surrounding the pupil in an eye) and processes it to develop a permanent Unique Verifiable Identity (UVI) for that person.[111]

Iris recognition is claimed to be one of the most secure and accurate biometric identification techniques currently available.[112] As IrisGuard reiterates, a single iris contains "more information than ten fingerprints combined".[113] A major advantage of iris recognition over other biometric techniques is that a person's iris does not change with time and consequently once-off enrolment ensures a person's unique identification over their entire lifetime, with no need for additional iris scans or captures. Iris identification also eliminates identity duplication, theft and trading. Furthermore, iris scans do not require direct contact and hence are hygienic, quick to perform and do not require large delicate devices.[114]

IrisGuard patented its iris recognition technology, EyePay®, which is an "innovative, secure financial delivery platform that replaces the use of traditional cards… and authorise(s) a fault-free transaction in less than three seconds".[115]

---

[109]IrisGuard, *Home Page*, www.irisguard.com, United Kingdom, accessed: 22 January 2020.

[110]United Nations, Sustainable Development Goals Knowledge Platform, *Sustainable Development Goal 16—Targets and Indicators*, https://sustainabledevelopment.un.org/sdg16, accessed: 21 February 2020.

[111]IrisGuard, *Iris Recognition*, www.irisguard.com/node/81, accessed: 22 January 2020.

[112]CH. Kalyani, *Various Biometric Authentication Techniques: A Review*, Journal of Biometrics and Biostatistics, 2017.

[113]Ibid.

[114]Ibid.

[115]IrisGuard, *About IrisGuard*, www.irisguard.com/node/29, accessed: 22 January 2020.

Fig. 2.14 Image showing the EyeCash® (an iris-enabled ATM), the EyeHood® (a portable iris imager), the EyePay® Phone® (an iris-enabled Android phone) and various iris-enabled POS devices (IrisGuard, *About IrisGuard*, www.irisguard.com/node/29, accessed: 22 January 2020)

Biometric technologies have been incorporated into numerous facets of everyday life, including: suspect identification in law enforcement, border security, civil administration (including voter identification at smart voting stations), patient and practitioner identification in the healthcare sector, as well as widespread global use for providing access control in offices, businesses, apartment blocks and state buildings.[116] IrisGuard focuses specifically on border control and financial inclusion applications. In partnership with UNHCR, IrisGuard has given 2.7 million refugees[117] a secure digital identity, which has played a crucial role in assisting UNHCR in operating refugee camps in Syria using the portable and durable EyeHood® iris scanner (see Fig. 2.14).[118]

IrisGuard's EyeCash® solution provides iris-enabled automated teller machines (ATMs) with the EyePay® Phone application allowing certain Android mobile

---

[116]Gemalto Company, *Biometrics: Authentication & Identification (definition, trends, use cases, laws and latest news—2020 review)*, www.gemalto.com/govt/inspired/biometrics, accessed: 22 January 2020.

[117]IrisGuard, *Success in Numbers*, www.irisguard.com/node/85, accessed: 22 January 2020.

[118]IrisGuard, *EyeHood*, www.irisguard.com/node/70, accessed: 22 January 2020.

phones and tablets to access the IrisGuard payment systems. It has also developed iris-enabled point-of-sale (POS) devices that facilitate card-less and cashless transactions.[119] Currently the company has dispensed over US$ 500 million in cash and processes approximately US$ 3 million of transactions every month in supermarkets.[120]

Moreover, these solutions also allow people to secure a receipt of salary and remit money abroad and, consequently, are ideal for migrant workers with no bank accounts, since the IrisGuard technologies ensure compliance with the Know Your Customer (KYC) and Anti-Money Laundering (AML) international regulations. It also enables people without a bank account to receive remittances from abroad as the transaction and beneficiary can be fully authenticated and the money withdrawn from an iris-enabled ATM.[121,122]

2.4.1.2 Turkey's Presidency's Communication Centre

Turkey's Directorate of Communications launched the Presidency's Communication Centre (CIMER) in 2015 with the aim to strengthen the "state-society relationship" within the country.[123] CIMER allows citizens to submit complaints, criticisms and compliments of public services via an online portal,[124] which are then forwarded to the relevant government entity and are monitored until a response is sent back to the citizen. In 2018 it was reported that more than 2.8 million online submissions were received, and officials reported that every one of the submissions received a response. The data from the portal are analysed to determine trends and generate reports that are presented to ministries and the Presidency to inform decisions and provide recourse.[125]

In 2019, CIMER was nominated as one of the "Champion Projects" at the World Summit on the Information Society (WSIS) Prizes under Category 3—Access to Information and Knowledge. The WSIS stated that the CIMER initiative assists with SDG 16—Peace, Justice and Strong Institutions, stressing the importance of the right to petition and for citizens to voice their opinions of public services. WSIS states that: "In e-government work, CIMER leads the creation of participatory public policies centred on the human by increasing the interaction between the state and the citizen.

---

[119]IrisGuard, *Technology*, www.irisguard.com/solutions2, accessed: 22 January 2020.

[120]IrisGuard, *Home Page*, www.irisguard.com, accessed: 22 January 2020.

[121]IrisGuard, *Remittances*, www.irisguard.com/node/15, accessed: 22 January 2020.

[122]IrisGuard, *Migrant Salary Assurance*, www.irisguard.com/node/14, accessed: 22 January 2020.

[123]Presidency of the Republic of Turkey, Directorate of Communications, *Presidency's Directorate of Communications receives the Best Project award with Cimer*, April 2019, www.iletisim.gov.tr/english/haberler/detay/presidency-s-directorate-of-communications-receives-the-best-project-award-with-cimer, accessed: 23 January 2020.

[124]Online portal: www.cimer.gov.tr/, accessed: 23 January 2020.

[125]Daily Sabah, *Turkish Presidency grabs international awards for public relations success*, July 2019, www.dailysabah.com/turkey/2019/07/12/turkish-presidency-grabs-international-awards-for-public-relations-success, accessed: 23 January 2020.

The opinions and suggestions of the public administration from all sections of the society to CIMER are analysed and they are [a] source of policy formation."[126]

Also in 2019 the Directorate of Communications launched the "I Have an Idea for My Country" initiative, which is similar to CIMER except that instead of complaints, the platform allows members of the public to submit suggestions and ideas about an array of issues (including health services, culture, energy production education and environmental sustainability) to the relevant public office. The ideas are evaluated by the relevant public office and, if deemed feasible, the authorities then contact the user that submitted the idea. In 2019, CIMER and the "I Have an Idea for My Country" project received the International Public Relations Association's (IPRA) Golden World Awards in Community Engagement and Public Affairs.[127]

2.4.1.3 Text4Baby Initiative, United States of America

The Text4Baby initiative is a PPP aimed at reducing the infant mortality rate in the USA and was launched in 2010 to target underserved pregnant women in low-income areas. In the first two years following its launch, more than 320,000 people enrolled with the national health text messaging service, which was the first of its kind in the USA.[128] Expectant mothers sign up to the service by simply sending a text message to the relevant number. After registering, users receive three text messages each week with health information relating to their specific stage of pregnancy up until their child is one year old.[129]

The information contained in the text messages covers numerous aspects of pregnancy, including: prenatal care, safe sleep, immunization, breastfeeding, nutrition, oral health, family violence, physical activity, safety, injury prevention, mental health, substance abuse, developmental milestones, labour and delivery.[130] An example of a Text4Baby message received during pregnancy is shown below:

> Free msg: If you have any signs of preterm labor—cramps, belly tightening, low back pain, bleeding, or watery, pink/brown discharge – call your Dr. straight away[131]

The content of the Text4Baby messages was developed by a team of professionals, including public health agencies at various levels (city, county, state and federal). Content development followed discussions with numerous target groups of pregnant

---

[126]WSIS, *WSIS Prizes Contest 2019 Nominee—Presidential Communication Center*, www.itu.int/net4/wsis/stocktaking/Prizes/2020/DetailsPopup/15429741913031580, accessed: 23 January 2020.

[127]Daily Sabah, *Turkish Presidency grabs international awards for public relations success*, July 2019.

[128]R. Whittaker et al., *Text4baby: Development and Implementation of a National Text Messaging Health Information Service*, American Journal of Public Health, Dec 2012, Vol 102.

[129]Text4Baby, *About*, www.text4baby.org/about/text4baby, accessed: 23 January 2020, Operated by Wellpass Inc., New York.

[130]Text4Baby, *Message Content Information*, https://partners.text4baby.org/index.php/about/message-content, accessed: 23 January 2020.

[131]Ibid.

women and new moms to ensure message content was aligned with the needs of the end-user.[132]

The structure of Text4Baby (a broad PPP that includes public offices, private companies, academic institutions, non-profit organizations and professional associations) is a key aspect that enabled the successful roll-out and uptake of the messaging services. Partnering with mobile service operators ensures the text messages can be sent for free thereby ensuring a free service for the end-user. Current partners include: Wellpass, CTIA Wireless Foundation, Grey Healthcare Group and Pfizer Inc. The government actors in the partnership include the US Department of Health and Human Services (HSS), White House Office of Science and Technology Policy, US Department of Defense Military Health System, US Department of Agriculture, US Consumer Product Safety Commission and the Social Security Administration.[133]

The high penetration of mobile phones globally offers a high-impact and low-cost method of reaching a large target audience and, as such, is an effective means for disseminating important healthcare information on a national level. Such systems seek to alter the behaviour of citizens by increasing their knowledge of a specific healthcare topic, which allows them to make informed decisions regarding their wellbeing. Mobile, health-based behavioural services have consequently realized a global uptake, but some have argued that not enough research has been conducted to determine their efficacy.[134] In 2012 a study was performed on the Text4Baby service, which involved a random pilot evaluation of 123 pregnant women using the Text4Baby service. The women were interviewed on their attitudes and behaviours relating to prenatal care before Text4Baby messages were delivered to the intervention group to develop a baseline. A follow-up interview was performed with each woman when their baby had an approximate gestational age of 28 weeks. The study revealed that the women who registered for Text4Baby exhibited an improved belief that they were knowledgeable and ready for motherhood. The study also noted varying levels of efficacy based on the education levels of the user.[135] Since 2012 Text4Baby has concluded numerous evaluation programmes and data analyses to ensure the maximum benefit of the service is realized. Some of the central findings of these evaluations are listed below:

1. The target population (i.e. underserved, pregnant women) is being reached, with the majority of participants residing in high-poverty areas, with over 52% coming from low-income households.[136]
2. Participants exhibit increased knowledge in critical topics relating to pregnancy and have improved beliefs around prenatal care, the risks of alcohol use, and the

---

[132]Ibid.

[133]Text4Baby, *Who Is Involved*, www.text4baby.org/about/who-is-involved, accessed: 23 January 2020.

[134]W. D. Evans et al., *Pilot evaluation of the Text4Baby mobile health program*, Research article licensed by BioMed Central Ltd, Washington, USA, 2012.

[135]Ibid.

[136]Text4Baby, *Data and Evaluation*, www.text4baby.org/about/data-and-evaluation, accessed: 23 January 2020.

importance of prenatal vitamins. Furthermore, a study performed by the George Washington University highlighted that Text4Baby participants were almost three times more likely to believe they were prepared for motherhood when compared to the control group.[137]

3. Behavioural change has been confirmed through other studies that revealed that Text4Baby mothers have improved glycemic control, lower postpartum alcohol consumption and improved rates of influenza vaccinations.[138]

2.4.1.4 Portugal's Citizen Shops and Citizen Spots

Portugal unveiled its first Citizen Shops in Lisbon in 1999 with the goal of augmenting the relationship between citizens and companies and the public administration offices. The Citizen Shops offer public services that combine various public and private entities in an endeavour to provide citizens with a "one-stop shop" for all their public service requirements, ranging from social security services, identification services, tax services and driving license services to private services such as electricity, water and television subscription services.[139]

The Citizen Shops provide extended working hours to ensure convenience and accessibility. The integrated range of services and extensive training of public servants ensures the processing of public services is a smooth and pleasant experience. The Citizen Shops have active Internet connections and connect to the online Citizen Portal,[140] which offers over 1500 public services made available by 580 different entities.[141] Public servants assist citizens in accessing and operating the online portal to overcome digital literacy barriers, thereby ensuring all citizens have access to the online public services. More than 153 million citizens used services offered by the Citizen Shops between 1999 and 2017.[142]

Since 2016, citizens have been able to automatically assess and comment on the level of service they received using QR codes or free text messages, which was introduced to ensure quality assurance. The Presidency for the Council of Ministers noted that the integrated nature of services offered at Citizen Shops facilitates

---

[137] Ibid.

[138] Ibid.

[139] Administrative Modernization Agency, *Citizen Shop*, Presidency of the Council of Ministers, Portugal, 2016, www.ama.gov.pt/web/english/citizen-shop, accessed: 23 January 2020.

[140] The Portuguese Citizen Portal, https://eportugal.gov.pt/cidadaos, provides more than 1500 public services via the Internet and is a result of merging the old Citizen Portal with the Business Portal and Entrepreneur Helpdesk to provide all public services through one site. A new addition to the portal is the "My Street" service which enables local civic participation through fault and occurrence reporting functionality along with suggested improvements and suggestions.

[141] Administrative Modernization Agency, *Citizen Portal*, Presidency of the Council of Ministers, Portugal, 2016, www.ama.gov.pt/web/english/citizen-portal, accessed: 23 January 2020.

[142] Administrative Modernization Agency, *Citizen Shop*, Presidency of the Council of Ministers, Portugal, 2016, www.ama.gov.pt/web/english/citizen-shop, accessed: 23 January 2020.

resource sharing and common infrastructure, which reduces the costs to the state and improves the efficiency of public administration duties.[143]

To extend the reach of online public services on offer, the Administrative Modernization Agency have added over 600 Citizen Spots to its network. Citizen Spots are located at Citizen Shops, local post offices and local government administration offices and provide approximately 200 public services of central and local administration, as well as private entities.[144] This model of piggy-backing on existing public infrastructure enables an extensive network that "allows serving the citizen better, quicker and closer, [thereby] promoting digital literacy through support in the provision of digital public services".[145]

The Observatory of Public Sector Innovation (OPSI) developed by the Organisation for Economic Co-operation and Development (OECD) "collects and analyses examples and shared experiences of public sector innovation to provide practical advice to countries on how to make innovations work".[146] It lauded the Citizen Shop project as a "very inspiring model of innovation in public service delivery" through which "citizen centricity has become the cornerstone of all new paradigms based on simplicity, efficiency and convenience".[147] However, it also cautioned that duplication of the model is only feasible if: (i) politicians at varying levels of government are directly involved and support the initiative, (ii) the service delivery is client-focused and encourages the "complaint-for-improvement" culture, (iii) effective partnerships between public institutions are possible, and (iv) public entities must agree on sharing information and integrating their databases.[148]

The OPSI noted that one of the key motivators of the success of the Citizen Shop initiative is the high quality standards expected of the shops, whose effectiveness is evaluated by three factors, namely: the number of visitors, the average waiting time before being helped, and the number of complaints.[149] Continued quality monitoring of shops results in a positive experience for the citizen and tailor-made, simplified service delivery options together with well-trained public servants have resulted in the success of the project.

---

[143]Ibid.

[144]Administrative Modernization Agency, *Citizen Stop*, Presidency of the Council of Ministers, Portugal, 2016, www.ama.gov.pt/web/english/citizen-spot, accessed: 23 January 2020.

[145]Ibid.

[146]Observatory of Public Sector Innovation, *The Observatory of Public Sector Innovation*, OECD, www.oecd.org/governance/observatory-public-sector-innovation/about/, accessed: 24 January 2020.

[147]Observatory of Public Sector Innovation, *Innovations—Citizen Shops—Lessons Learned*, www.oecd.org/governance/observatory-public-sector-innovation/innovations/page/citizenshops.htm#tab_lessons, accessed: 24 January 2020.

[148]Ibid.

[149]Observatory of Public Sector Innovation, *Innovations—Citizen Shops—Results*, www.oecd.org/governance/observatory-public-sector-innovation/innovations/page/citizenshops.htm#tab_results, accessed: 24 January 2020.

2.4.1.5 X-Road—An Open Source Data Exchange Layer

Providing a broad spectrum of online public services is only possible if the databases and ICT infrastructures of all government departments involved are interoperable. Ensuring interoperability across numerous different information systems can be a difficult and expensive task, especially if such systems have been developed independently over many years. This is often one of the main contributors to change resistance from ministers and other decision makers when new e-government applications are proposed.

To overcome this problem, Estonia's Information System Authority developed X-Road, "an open source data exchange layer solution that enables organizations to exchange information over the internet".[150] The centrally managed distributed data exchange layer not only provides interoperability between data exchange parties but also ensures the security, confidentiality and integrity of all data transfers through encryption, authentication and digital signatures.[151] Furthermore, its structure and architecture nullify traditional attacking vectors, thereby ensuring security.[152] It also provides functionality for writing to multiple information systems simultaneously, transmitting large data sets and search functionality across several information systems concurrently.[153] Other features provided by X-Road include: address management, message routing, access rights management, organization level authentication, machine level authentication, transport level encryption, time-stamping, digital signing of messages, logging and error handling.[154]

As such, X-Road provides governments with a critical tool for linking their various departments and respective information systems (with varying services and database structures), without redesigning system architectures or replacing hardware, as illustrated in Fig. 2.15. The X-Road system and its code is open source, released under the MIT open source license and is available for free to any organization or individual.[155]

National deployment of X-Road in Estonia began in December 2001. The Estonian data exchange layer, subsequently renamed X-Tee, has 156 public sector institutions, 503 private institutions and enterprises, 1,189 interfaced information systems and

---

[150]Nordic Institute for Interoperability Solutions, *X-Road—The free and open source data exchange layer software*, Estonia, https://x-road.global/?__hstc=120471575.3f60f35c1716c8d1df fc3e6e3ca56f7a.1582635703187.1582635703187.1582709064461.2&__hssc=120471575.1.158 2709064461&hsfp=627436306, accessed: 3 February 2020.

[151]Ibid.

[152]Estonian Information System Authority, *X-tee Fact Sheet*, www.x-tee.ee/factsheets/EE/#eng, accessed 28 February 2020.

[153]E-Estonia, *Interoperability Services*, https://e-estonia.com/solutions/interoperability-services/x-road/, accessed: 3 February 2020.

[154]Nordic Institute for Interoperability Solutions, *X-Road—The free and open source data exchange layer software*, Estonia, https://x-road.global/?__hstc=120471575.3f60f35c1716c8d1df fc3e6e3ca56f7a.1582635703187.1582635703187.1582709064461.2&__hssc=120471575.1.158 2709064461&hsfp=627436306, accessed: 3 February 2020.

[155]Ibid.

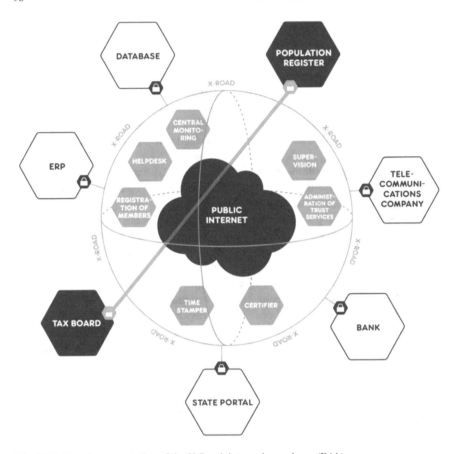

Fig. 2.15 Visual representation of the X-Road data exchange layer (Ibid.)

approximately 52,000 organizations as indirect users of X-Tee services.[156] Through the use of X-Tee, Estonia has managed to ensure that 99% of all its state services are available online.[157] In 2019, there were over 1.3 billion requests on X-Tee, with 3% of those requests submitted by citizens. Assuming each citizen request saves 15 min, X-Tee resulted in a staggering saving of 1,134 working years in 2019 alone.[158]

---

[156]Estonian Information System Authority, *X-tee Fact Sheet*, www.x-tee.ee/factsheets/EE/#eng, accessed 28 February 2020.

[157]E-Estonia, *Interoperability Services*, https://e-estonia.com/solutions/interoperability-services/x-road/, accessed: 3 February 2020.

[158]Estonian Information System Authority, *X-tee Fact Sheet*, www.x-tee.ee/factsheets/EE/#eng, accessed 28 February 2020.

Since its success in Estonia and Finland, the X-Road data exchange layer has been implemented[159] in the following countries: Azerbaijan, Argentina, Columbia, El Salvador, Faroe Islands, Germany, Iceland, Japan, Kyrgyzstan, Palestine and Vietnam.[160]

Additionally, X-Road implementation is in the consultation or planning phase in: Brazil, Cambodia, Canada, Cayman Islands, Denmark, Dominican Republic, Guatemala, India, Israel, Kazakhstan, Kurdistan, Madagascar, Mexico, Norway, Panama, Scotland, Spain, Sweden, Uruguay and Venezuela.[161]

An interesting application of X-Road[162] in Estonia is the use of the data exchange layer to deliver and consume energy in a more sustainable way using Elering's smart grid platform. Elering, the independent electricity and gas transmission operator in Estonia created Estfeed—a data access and exchange platform based on X-Road, which enables a smart grid by connecting the needs of energy consumers (data owners), energy providers (data users) and smart meter hubs (data providers). The data exchange layer allows for effective and informative analysis of consumption, production and transmission data to optimize the efficiency and reliability of the energy grid.[163] Estfeed is aiming to become the energy data platform used across the European Union (EU) and is currently being refined according to the EU General Data Protection Regulation (GDPR) guidelines[164] and the Clean Energy for All Europeans Package[165] (an updated energy policy framework focused on aiding countries transitioning from fossil fuels towards cleaner energy).[166] Figure 2.16 provides a visual representation of the data flows within the Estfeed system.

---

[159]Implementation is based on the X-Road core open source codebase and has full protocol-level compatibility with the official X-Road core according to the Nordic Institute for Interoperability Solutions.

[160]Nordic Institute for Interoperability Solutions, *X-Road World Map—Implementation projects based on the open source codebase with full protocol-level compatibility*, https://x-road.global/xroad-world-map, accessed: 28 February 2020.

[161]Ibid.

[162]To read about other X-Road case studies visit: *Ensuring interoperability in water services and waste management information systems in Helsinki*, https://x-road.global/environmental-services, *Providing information exchange between the national tax boards of Estonia and Finland*, https://x-road.global/case-study-tax-boards, *Digitalising public services for citizens, industries and the public sector in the Faroe Islands*, https://x-road.global/empowering-citizens-and-business, *Cross-border interoperability and information exchange between the national business registers of Estonia and Finland*, https://x-road.global/case-study-the-business-registers-of-estonia-and-finland.

[163]Nordic Institute for Interoperability Solutions, *Case Study—Access to electricity and gas smart meter data in Estonia powered by X-Road technology*, https://x-road.global/access-to-electricity-and-gas-smart-meter-data-in-estonia, accessed: 3 February 2020.

[164]Proton Technologies AG, *Complete guide to GDPR* compliance, 2020, https://gdpr.eu/, accessed: 28 February 2020.

[165]European Commission, *Clean energy for all Europeans* package, https://ec.europa.eu/energy/en/topics/energy-strategy/clean-energy-all-europeans, published: October 2017, updated: February 2020, accessed: 28 February 2020.

[166]Nordic Institute for Interoperability Solutions, *Case Study—Access to electricity and gas smart meter data in Estonia powered by X-Road technology*, https://x-road.global/access-to-electricity-and-gas-smart-meter-data-in-estonia, accessed: 3 February 2020.

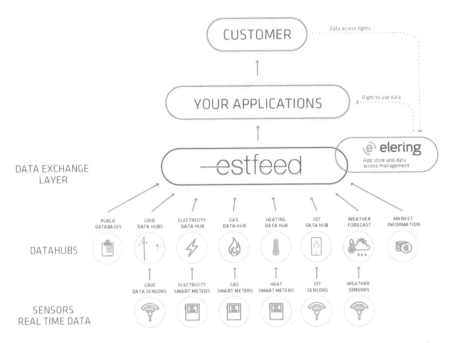

Fig. 2.16 Visual representation of the data flows within the Estfeed system (Nordic Institute for Interoperability Solutions, *Case Study—Access to electricity and gas smart meter data in Estonia powered by X-Road technology*, https://x-road.global/access-to-electricity-and-gas-smart-meter-data-in-estonia, accessed: 3 February 2020)

2.4.2   Exemplary E-Government Applications in the Developing World

This section examines four examples of e-government applications in developing countries. First, Vodafone's M-Pesa, Africa's largest mobile money service promoting financial inclusion without the need for a formal bank account, is discussed. Zipline, a California start-up that is delivering urgent medical supplies in rural Africa through the use of autonomous drones is then analysed. Next, the successful transformation and development of Ghana's ICT infrastructure and economy through the e-Ghana and e-Transformation projects is assessed. The two projects funded by the World Bank provided an enabling environment for e-government applications within Ghana. Lastly, India's Aadhaar project is presented. Aadhaar is the world's largest digital biometric identification system, which covers more than 99% of India's adult population. The numerous impacts of the project, specifically those resulting in financial inclusion and the optimization and extension of social subsidies and welfare schemes, are examined.

2.4.2.1 M-Pesa—Providing Financial Services Through Mobile Devices

M-Pesa was developed by Vodafone in the United Kingdom after obtaining a grant from the United Kingdom's Department for International Development designed for providing financial services to the millions of unbanked people in developing countries.[167] Kenya was selected as the first country to trial the system, where in 2006 only 14% of the population had formal bank accounts.[168] Initial development and testing focused on microfinance loans using mobile phones and was trialled in Thika, Kenya. The trial revealed one of the largest financial challenges in Kenya and the rest of Africa, namely that breadwinners typically travel to urban areas to source work and then send money back home to their families based in rural areas. As such, Vodafone re-engineered the system to provide easy, affordable mobile money transfers based on SMS messages and local agents, such as retail outlets and airtime resellers. The new system, M-Pesa, allowed individuals to deposit cash at a local agent and then use SMS messages to convert the money into electronic funds linked to their SIM cards. Then, using SMS messages, the electronic funds could be transferred to anyone in the country with a mobile phone who could visit a local agent and withdraw a cash equivalent. M-Pesa was launched by Safaricom, Vodafone's Kenyan associate, in March 2007 and by the end of its first year it had serviced 1.2 million customers.[169]

Today, M-Pesa has over 37 million active customers with almost 400,000 active agents and is Africa's largest mobile money service. Last year alone, there were over 11 billion transactions on the M-Pesa system, which is operating in the Democratic Republic of Congo, Egypt, Ghana, Kenya, Lesotho, Mozambique and Tanzania.[170] M-Pesa has grown to include additional financial services that include merchant payment services, microloans and saving accounts with local banks.[171]

M-Pesa also allows businesses to pay employees without bank accounts, to make stock purchases, and to collect payments from customers without the need for cash.

---

[167]M. Joseph (Vodafone Group's Director of Mobile Money), *M-Pesa: The story of how it was created*, March 2017, www.vodafone.com/perspectives/blog/m-pesa-created, accessed: 10 February 2020. For additional information see: D. Lindgren, "Space-Based Financial Services and Their Potential for Supporting Displaced Persons", in *Embedding Space in African Society: The United Nations Sustainable Development Goals 2030 Supported by Space Applications*, ed. A. Froehlich (Cham: Springer, 2019), 93–104.

[168]G. Obulutsa, *M-Pesa has completely changed Kenyans' access to financial services, this is how...*, CNBC Africa, April 2019, www.cnbcafrica.com/news/east-africa/2019/04/03/m-pesa-has-completely-changed-kenyans-access-to-financial-services-this-is-how/, accessed: 10 February 2020.

[169]M. Joseph (Vodafone Group's Director of Mobile Money), *M-Pesa: The story of how it was created*, March 2017, www.vodafone.com/perspectives/blog/m-pesa-created, accessed: 10 February 2020.

[170]Vodafone Group, *What is M-Pesa*, 2020, www.vodafone.com/what-we-do/services/m-pesa, accessed: 10 February 2020.

[171]G. Obulutsa, *M-Pesa has completely changed Kenyans' access to financial services, this is how...*, CNBC Africa, April 2019, www.cnbcafrica.com/news/east-africa/2019/04/03/m-pesa-has-completely-changed-kenyans-access-to-financial-services-this-is-how/, accessed: 10 February 2020.

Similarly, the service enables governments to collect taxes and provide social security payments to its citizens. This also applies to NGOs and other charity organisations wishing to pay money to beneficiaries with no bank accounts. In Tanzania the Comprehensive Community Based Rehabilitation NGO uses the M-Pesa platform to pay for patients' travel to hospitals. Similarly, M-Kopa Solar, a company selling affordable solar electricity systems in Kenya and Tanzania, allows M-Pesa users to pay off solar installations with daily contributions so that after one year the installation is paid off and fully owned by the household.[172]

A survey conducted by the Kenyan central bank revealed that in 2019, 83% of Kenya's population had access to formal financial services, highlighting how mobile devices and mobile financial services have directly contributed to this increase.[173] M-Pesa is not the only mobile financial service in Kenya, Airtel Kenya, Telkom Kenya Ltd and Equity Group (through their subsidiary Equitel) offer similar services using mobile devices. The Communications Authority of Kenya revealed that at the end of 2018 there were 31.6 million active users of mobile money transfer services, constituting almost 64% of the country's population. M-Pesa controlled the majority of the market, with 25.57 million active yearly users in Kenya, while Airtel had 3.77 million.[174]

M-Pesa is an example of leveraging mobile technology to overcome certain challenges in developing countries to provide services to disjointed communities with poor infrastructure and promoting inclusion. It also highlights how providing such vital services to underserved communities can be profitable for private companies. In fact, in the financial year ending in March 2019, M-Pesa had generated US$ 743 million of revenue for Safaricom.[175] Figure 2.17 depicts how M-Pesa's annual revenue has increased year on year as its user base increases.

However, when developing and establishing new solutions in developing countries, it is critical to be cognizant of the local socio-economic and legislative climate, which can vary greatly between countries in the same region. This was highlighted by the failure of M-Pesa in South Africa. After achieving great success in other African countries, M-Pesa was launched in 2010 in South Africa through partnering with the local bank, Nedbank. M-Pesa's target was to foster a customer base of 10 million users in its first three years.[176] However, after six years the service only achieved 76,000 users and in June 2016 the service was shut-down entirely in South Africa.[177]

---

[172]Vodafone Group, *What is M-Pesa*, 2020, www.vodafone.com/what-we-do/services/m-pesa, accessed: 10 February 2020.

[173]G. Obulutsa, *M-Pesa has completely changed Kenyans' access to financial services, this is how...*, CNBC Africa, April 2019, www.cnbcafrica.com/news/east-africa/2019/04/03/m-pesa-has-completely-changed-kenyans-access-to-financial-services-this-is-how/, accessed: 10 February 2020.

[174]Ibid.

[175]R. Gupta, *Safaricom and Vodacom to team up to buy M-Pesa rights*, Market Realist, July 2019, https://articles2.marketrealist.com/2019/07/safaricom-vodacom/#, accessed: 10 February 2020.

[176]L. Mbele, *Why M-Pesa failed in South Africa*, BBC Africa business Report, Johannesburg, South Africa, May 2016, www.bbc.com/news/world-africa-36260348, accessed: 10 February 2020.

[177]G. van Zyl, *Why Vodacom M-Pesa has flopped in SA*, fin24, May 2019, www.fin24.com/Tech/Companies/why-vodacom-m-pesa-has-flopped-in-sa-20160509, accessed: 10 February 2020.

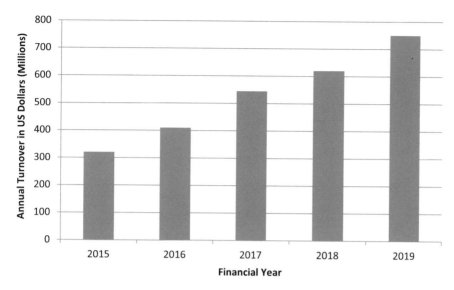

Fig. 2.17 M-Pesa annual revenue over the past five years (Ibid.)

One of the key reasons for the failure of the service in South Africa was the absence of a need for new services to increase financial inclusion. In 2016, approximately 75% of South Africans had formal bank accounts. Moreover, South Africa already had a well-established money transfer market through various service providers including local retail chains and formal banks (that already offered mobile money transfers).[178]

2.4.2.2 Using Autonomous Drones to Deliver Medical Supplies in Rwanda

Logistical operations in developing countries are often complex and challenging as a result of dispersed populations in rural areas separated by large geographic distances. This coupled with poor road infrastructure and challenging terrains often results in costly delays in deliveries. Such delays pose a threat to the delivery of critical medical supplies that require a constant cold-chain, with no interruptions.

This is especially true in Rwanda, where the majority of the country's 12.6 million population live in rural communities dispersed around the country. The capital of Rwanda, Kigali, has a population of just 860,000 people, and the population of the second largest city in the country, Butare, is under 90,000.[179] The widely distributed

---

[178]Ibid.

[179]S. Fleming, *in Rwanda, high-speed drones are delivering blood to remote communities*, World Economic Forum, December 2018, www.weforum.org/agenda/2018/12/in-rwanda-drones-are-del ivering-blood-to-remote-communities/, accessed: 10 February 2020.

Fig. 2.18 A Zipline employee packing a package for delivery into a drone (Zipline YouTube Video, *Zipline's World Class Drone Safety Features*, uploaded on 28 February 2019, https://youtu.be/QTn Bm3rXmss, accessed: 10 February 2020)

nature of the country's population makes providing healthcare to all citizens a challenging task, especially for delivering blood used for transfusions, which requires a constant cold-chain throughout the delivery process. This problem is exacerbated by the fact that only 25% of the country's roads are paved or surfaced. According to Rwanda's Minister of Health, the leading cause of maternal deaths in Rwanda is postpartum haemorrhaging, which has been confirmed as the leading cause of maternal deaths around the world by the World Health Organisation (WHO). As such, the availability of fresh blood supplies is paramount to providing adequate healthcare in hospitals and clinics.[180]

California-based technology start-up, Zipline, founded in 2014, has designed an uncrewed and autonomous system (UAS) that uses drones to deliver blood supplies to hospitals and clinics around Rwanda.[181] Clinics request a blood delivery using SMS messages or WhatsApp messages from a central storage facility, specifying the blood type and amount required. The blood is then packaged and secured on a drone (see Fig. 2.18) that travels to the clinic autonomously and releases the delivery with a parachute, as shown in Fig. 2.19. The drop-off point is automatically calculated by

---

[180]Ibid.

[181]Zipline, *Our Mission*, https://flyzipline.com/company/, accessed: 10 February 2020.

Fig. 2.19 Zipline drone delivering a package equipped with a parachute (Zipline, *How it works*, https://flyzipline.com/how-it-works/, accessed: 10 February 2020)

the drone to account for local weather conditions. By 2018, Zipline was responsible for more than 35% of the deliveries for Rwanda's national blood transfusions outside the capital and supplied 21 remote clinics with fresh blood.[182] As of February 2020, the company had performed more than 31,000 commercial deliveries and has flown more than 2 million kilometres with their drones to customers.[183,184]

The drones reduce delivery time of fresh blood from a few days by vehicle, to just 20 minutes. These fast delivery times reduce overstocking of clinics and wasted blood from expiries. Furthermore, the quick delivery time translates to no interruptions in the cold-chain and this lessens the amount of blood that has to be discarded. The drone service also results in fewer patient referrals, as clinics no longer have to turn patients away due to no stock—they can simply order new blood supplies that will arrive within 20 minutes.[185]

---

[182]S. Fleming, *in Rwanda, high-speed drones are delivering blood to remote communities*, World Economic Forum, December 2018, www.weforum.org/agenda/2018/12/in-rwanda-drones-are-delivering-blood-to-remote-communities/, accessed: 10 February 2020.

[183]Zipline, *Home page*, https://flyzipline.com/, accessed: 10 February 2020.

[184]Zipline, *How it works*, https://flyzipline.com/how-it-works/, accessed: 10 February 2020.

[185]Zipline, *Solutions—Global public health*, https://flyzipline.com/solutions/global-public-health/, accessed: 10 February 2020.

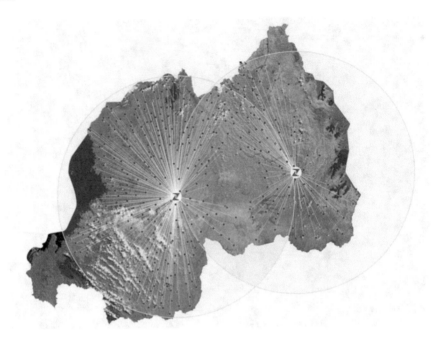

Fig. 2.20 Map of Zipline services in Rwanda from two distribution centres (Zipline, *Home page*, https://flyzipline.com/, accessed: 10 February 2020)

To ensure the safety of citizens on the ground, the drones are completely redundant, with each drone having two motors, two power systems, two communication modules, and two navigation systems. Thus, if one fails the other module ensures uninterrupted operations. Each drone is also equipped with a safety parachute that deploys in the event of a drone being critically damaged and no longer able to fly. This ensures that the drone will land safely and eliminates the chance of injuring people on the ground.[186]

The fixed-wing drones can support payloads of up to 1.75 kg and run on lithium-ion batteries with a capacity exceeding 160 km. The drones are launched using a physical Zipline (the inspiration behind the company's name) and have a cruising speed of 100 km/h. They can travel through adverse weather both at day and night. The drones are stored, maintained and launched from central distribution centres that also act as warehousing facilities. Each distribution centre can perform hundreds of deliveries per day, moving almost one ton of inventory, and has a service radius of 85 km (covering a service area of more than 25,000 km^2).[187] Figure 2.20 shows how just two Zipline centres in Rwanda are sufficient to provide service to the entire country.

---

[186]Zipline YouTube Video, *Zipline's World Class Drone Safety Features*, uploaded on 28 February 2019, https://youtu.be/QTnBm3rXmss, accessed: 10 February 2020.

[187]Ibid.

A crucial element contributing to the success of Zipline was the close collaboration with Rwanda's Civil Aviation Authority (CAA) throughout the development process. This collaboration ensured the developed drones were in accordance with the requirements of the CAA, saving time during the development process. Zipline distribution centres are in direct contact with Rwanda's CAA and inform them of all planned flights. Additionally, the CAA can view the location of all active drones at any time, allowing them to track the drones as necessary.[188]

Following their success in Rwanda, Zipline launched a similar program in Ghana in 2019. Using a total of 30 drones, operating out of four distribution centres to deliver vaccines, blood and life-saving medications, Zipline provided services to over 2,000 health facilities across the country. The CEO of Zipline confirmed that with this infrastructure, they expect to perform approximately 600 flights each day in Ghana alone, servicing a total of 12 million people. This makes the program in Ghana the largest drone delivery service in the world.[189]

Zipline is currently exploring using their drone delivery service in other applications, including expanding access to clinical trials, on-demand delivery for pharmacies and health systems, and distributed resupply missions in defense and disaster scenarios.[190]

However, comprehensive national legislation still remains a hurdle in rolling out automated drone delivery services in most countries, especially in developed nations. As such, Africa has become an area where commercial drone delivery services and associated regulatory frameworks are being piloted. Certain African countries have been quick to update their legislation in this regard, with South Africa releasing legislation for the training and licensing of commercial drone pilots and Malawi developing a drone testing corridor for humanitarian applications in partnership with UNICEF.[191] UNICEF has launched similar corridors in Kazakhstan, Sierra Leone and Vanuatu.[192]

2.4.2.3 Ghana's ICT and E-Government Transformation

Ghana is an example of how targeted development in the ICT sector, leveraging PPPs, can increase government's ability to provide e-governance solutions as well as create employment opportunities and boost the economy. Ghana achieved this

---

[188] Ibid.

[189] J. Bright, *Drone delivery startup Zipline launches UAV medical program in Ghana*, Tech Crunch, Verizon Media, April 2019, https://techcrunch.com/2019/04/24/drone-delivery-startup-zipline-launches-uav-medical-program-in-ghana/, accessed: 10 February 2020.

[190] Zipline, *How it works*, https://flyzipline.com/how-it-works/, accessed: 10 February 2020.

[191] J. Bright, *Drone delivery startup Zipline launches UAV medical program in Ghana*, Tech Crunch, Verizon Media, April 2019, https://techcrunch.com/2019/04/24/drone-delivery-startup-zipline-launches-uav-medical-program-in-ghana/, accessed: 10 February 2020.

[192] UNICEF, *UNICEF expands network of drone testing corridors*, New York, April 2019, www.unicef.org/press-releases/unicef-expands-network-drone-testing-corridors, accessed: 10 February 2020.

through two projects funded by the World Bank, specifically the e-Ghana project, which commenced in November 2006 and ended in December 2014,[193] and the e-Transform project, which commenced in 2013 and will terminate at the end of December 2020.[194] A third project, the Ghana Economic Transformation Project, began in 2019 and will build on the successes of the previous two projects, aiming to promote private investments and growth in non-resource based sectors to reduce the volatility and non-inclusive nature of commodities-based economies. The Economic Transformation Project will run until November 2025.[195]

The success of the e-Ghana and e-Transform projects is observable in Ghana's improvements in the UN E-Government Surveys over the past ten years. In 2012, Ghana ranked 145th in the world in the survey. In 2018 Ghana climbed 44 places and achieved a ranking of 101 globally and was the fifth highest ranked country in Africa. Figure 2.21 compares Ghana's EGDI, OSI, TII and HIS scores obtained in the 2012 survey with those obtained in the 2018 survey. Ghana achieved a 70% increase in its overall EGDI rating, which rose from 0.316 in 2012 to 0.539 in 2018.[196,197]

The two categories where Ghana improved most are the OSI and TII categories, which reflect the success of the two aforementioned projects whose objectives included increasing the number of online government services and developing the country's ICT infrastructure. The OSI rating more than doubled from 0.301 in 2012 to 0.694 in 2018 and the TII rating more than tripled, which is a remarkable accomplishment considering the relatively short time frame of six years.

The e-Ghana project commenced after the Government of Ghana (GoG) approached the World Bank Group to assist in increasing the ICT economy in the country with the intention of increasing economic output, attracting foreign investment and creating new employment opportunities. The two core objectives of the e-Ghana project were to develop the information technology enabled services (ITES) sector and increase e-government applications in an effort to increase the efficiency, accountability, responsiveness and transparency of public administrative duties. PPPs and ICT infrastructure were identified as key enablers supporting these goals.[198]

---

[193]Independent Evaluation Group (IEG), *e-Ghana Project—Project Performance Assessment Report*, World Bank Group, Report No. 108359, 2016.

[194]The World Bank, *Ghana e-Transform Project—additional financing and restructuring*, https://pro jects.worldbank.org/en/projects-operations/project-detail/P144140#results, accessed: 14 February 2020.

[195]The World Bank, *Combined project information documents/integrates safeguards datasheet (PID/ISDS)*, Ghana Economic Transformation Program, April 2019, http://documents.worldbank. org/curated/en/872011556213034939/pdf/Project-Information-Document-Integrated-Safeguards-Data-Sheet-Ghana-Economic-Transformation-Project-P166539.pdf, accessed: 14 February 2020.

[196]United Nations Department of Economic and Social Affairs, *E-Government Survey 2012—E-Government for the People*, New York, 2012.

[197]United Nations Department of Economic and Social Affairs, *E-Government Survey 2018—Gearing E-Government to Support Transformation towards Sustainable and Resilient Societies*, New York, 2018.

[198]World Bank Group, *e-Ghana Project Information Document*, Report No. 36790, Washington, USA, 2006.

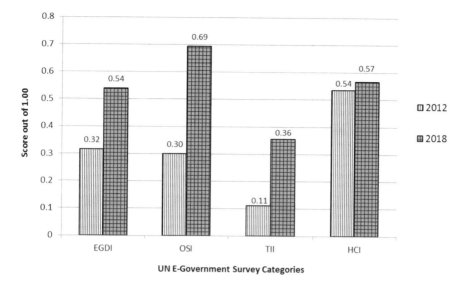

Fig. 2.21 Ghana's scores in the UN *E-Government Survey* in 2012 and 2018 (Image created from data contained within the *UN E-Government Survey 2012*, https://publicadministration.un. org/publications/content/PDFs/E-Library%20Archives/UN%20E-Government%20Survey%20s eries/UN%20E-Government%20Survey%202012.pdf, accessed: 29 February 2020, and *UN E-Government Survey 2018*, https://publicadministration.un.org/Portals/1/Images/E-Government% 20Survey%202018_FINAL%20for%20web.pdf, accessed: 29 February 2020)

The implementing agency within Ghana for the projects is the Ghanaian Ministry of Communications, which "is responsible for national telecom policy, to facilitate the strategic development and application of information and communication technologies for socio-economic growth".[199] The e-Ghana project consisted of four core components:

1. Create an enabling environment for the development of the ICT sector through[200]:

    (a) Supporting the Ministry of Communications in their operational and implementation activities as well as providing capacity-building for such roles,
    (b) Developing and implementing the required policies and legal frameworks, including a national ITES policy and other regulatory instruments and checks aimed at improving the quality of telecommunication services and reducing their costs,
    (c) Preparing key ICT legislations regarding cybersecurity, intellectual property, data protection and consumer protection,
    (d) Investigating different options for realizing sustainable community information centres.

---

[199]Ibid.
[200]Ibid.

2. Support local ICT businesses and the ITES industry within Ghana by[201]:

    (a) Developing ITES human capacity by establishing standard skill set require-
        ments, curricula and accreditation mechanisms to provide training programs
        and grants,
    (b) Implementing investment strategies to attract international investors in the
        business process offshoring industry,
    (c) Supporting local private business in the ICT sector through provision of
        facilities to provide international certifications, developing ITES business
        incubators, and nurturing existing ITES businesses in Ghana by providing
        operational support as necessary.

3. Leverage ICT to increase e-government applications and online services by[202]:

    (a) Developing and implementing standards governing interoperability and IT
        architectures for government information systems,
    (b) Establishing a country-wide wide-area network (WAN) to connect key
        government ministries, departments and agencies, enabling information
        sharing,
    (c) Establishing national data centres, payment gateways and authentica-
        tion systems to facilitate delivery of e-government systems and improve
        efficiencies of public administration duties,
    (d) Training technical staff and information officers of key ministries, depart-
        ments and agencies,
    (e) Investigating PPPs to support implementation of electronic applications for
        e-government services, including the Ghana Revenue Authority.

4. Develop the Ghana Integrated Financial Management Information System
    (GIFMIS) used to prepare budgets, generate reports and perform accounting
    of all GoG funds.[203]

All four components of the project were successful overall, and an enabling environment for the ICT and ITES sector was created. By the end of the project, a total
of 8,700 new jobs were created across the two sectors, which exceeded the goal
of 7,000 (with 54.3% of the new jobs being held by women).[204] Furthermore, the
new legislation and policies ensured continued foreign investment that assisted in
growing the local ICT sector and enabled an average annual growth rate of 23.3%
over the duration of the project in the sector. The economic benefit of this growth
was reflected in Ghana's GDP, which exhibited an 8.1% average annual growth rate

---

[201] Ibid.

[202] Ibid.

[203] Ibid.

[204] The World Bank, *Implementation Completion and Results Report of the e-Ghana Project*, World
Bank Group, Report No. ICR00003288, 2015.

over the same period (compared to the baseline value of 5.8% before the project commenced).[205]

Despite the World Bank's Independent Evaluation Group noting a substantial impact of the project in terms of increasing government efficiency and transparency, a lack of quantifiable metrics in this regard was noted. Nonetheless, automating the processes of the Ghana Revenue Authority resulted in[206]:

- 87,900 new automated business registrations,
- 425,305 tax-payer identification number registrations and
- An estimated 400,000 new tax-payers.

Furthermore, the revenue collected via the automated tax collection service (named the Total Revenue Integrated Processing System, or TRIPS) increased from zero in 2012 to almost 62% of the total tax collection in 2015. Amazingly, the amount of tax collected in 2015 was three times more than that collected in 2010.[207]

The project established over 14 different online public services, including online marriage registration, company registration, marriage licenses, certified copies of birth certificates and criminal background checks.[208] At the end of the project, 100% of new business registrations were performed using e-government applications and approximately 60% of all new marriage licenses were completed using the online platform.[209] However, the other e-services exhibited a very low usage rate of less than 10% which was attributed to a lack of public awareness combined with low Internet access and poor connectivity.[210]

GIFMIS was also deemed a success. By the end of the project, all 33 ministries in Ghana were connected to the system and actively using it for accounting, forecasting and reporting. In 2015 GIFMIS was being used for 66% of the total public expenditure.[211]

The total cost of the project was US$ 113 million, with US$ 80.25 million coming from the World Bank's International Development Association (IDA) resources and US$ 27 million co-financed by the EU and the UK Department for International Development (DFID). Five PPPs were established for the provision of services supporting online public transactions and systems.[212]

The e-Transform project was approved in 2013 and endeavours to build on the success of the e-Ghana project to "improve the efficiency and coverage of government

---

[205]Independent Evaluation Group (IEG), *e-Ghana Project—Project Performance Assessment Report*, World Bank Group, Report No. 108359, 2016.

[206]Ibid.

[207]Ibid.

[208]Ibid.

[209]The World Bank, *Implementation Completion and Results Report of the e-Ghana Project*, World Bank Group, Report No. ICR00003288, 2015.

[210]Independent Evaluation Group (IEG), *e-Ghana Project—Project Performance Assessment Report*, World Bank Group, Report No. 108359, 2016.

[211]Ibid.

[212]Ibid.

service delivery using ICT".[213] The core objectives of the project include: reducing the average number of days to perform government services, providing access to online educational portals to 5,000 teachers and students, developing seven new e-government applications (including e-Justice, e-Immigration and e-Procurement), and providing 16 new online e-services to the public.[214]

Key sectors that will be transformed by the project include: the public administration, health, education and judiciary sectors. The project consists of four components[215]:

1. Component 1: Enabling Environment for Electronic Government
2. Component 2: Common Services and Infrastructure for Electronic Government
3. Component 3: Scale Up of e-Services and Applications
4. Component 4: Project Management Support.

The first component will focus on developing new (and revising existing) policies, laws, regulations and frameworks aimed at providing a transparent and secure environment for promoting the provision of electronic services. The component consists of three sub-components, specifically: (i) ensuring open government data is securely stored and protected, (ii) supporting entrepreneurship and job creation and (iii) capacity building at an institutional level for regulatory bodies.[216] The objectives for this component also include establishing three new technology hubs and three new certification authorities in Ghana, as well as making 500 government datasets available online.[217]

The second component focuses on digitization and connectivity of public agencies within Ghana in support of e-governance. Key objectives include digitizing over 40 million government records, including 14 million birth and death records and 5 million records in beneficiary institutions for e-Applications (including the judicial service and immigration services). Additional objectives include providing 50 schools and universities with high-speed Internet connectivity.[218]

The third component seeks to improve service delivery in the health and education sectors through scaling up e-Services and applications and includes developing

---

[213]The World Bank, *e-Transform Ghana Implementation Status & Results Report*, Sequence No. 11, World Bank Group, October 2019, http://documents.worldbank.org/curated/en/616411572 406052343/Disclosable-Version-of-the-ISR-GH-eTransform-Ghana-P144140-Sequence-No-11, accessed: 28 February 2020.

[214]Ibid.

[215]Ministry of Communications—Republic of Ghana, *E-Transform Project*, Accra, Ghana, www. moc.gov.gh/e-transform-project, accessed: 14 February 2020.

[216]The World Bank, *Ghana e-Transform Project—additional financing and restructuring*, https://pro jects.worldbank.org/en/projects-operations/project-detail/P144140#results, accessed: 14 February 2020.

[217]The World Bank, *e-Transform Ghana Implementation Status & Results Report*, Sequence No. 11, World Bank Group, October 2019, http://documents.worldbank.org/curated/en/616411572 406052343/Disclosable-Version-of-the-ISR-GH-eTransform-Ghana-P144140-Sequence-No-11, accessed: 28 February 2020.

[218]Ibid.

a strategic plan for an integrated e-Health system. The objectives in this component consist of training 600 teachers through Internet-aided courses and training 1,000 people in e-Services. As of October 2019, both of these objectives were exceeded, with almost 1,400 teachers being trained through Internet-aided education and over 2,000 people trained in e-Services. Further, as of December 2016, almost 95,000 students and teachers benefited from improved online access through new infrastructure at schools and universities established under the project.[219]

The e-Transform project will close in December 2020. The World Bank approved the Ghana Economic Transformation Project in 2019, which will follow on from the e-Transform project and focuses on improving the ease of doing business within the country in an attempt to secure private investments.[220]

Ghana provides an illustration of how countries can obtain foreign investment opportunities afforded by international development agencies to improve their ICT infrastructure, and leverage it to create new job opportunities, increase their economic standing and simultaneously improve their online public presence and e-governance. The case studies of the e-Ghana and e-Transformation projects highlight the importance of establishing a transparent regulatory framework and legislation to ensure an enabling environment for developments in the ICT sector. The standards, laws, regulations and policies established under the e-Ghana project were critical to securing funding and boosting private investment within the ICT sector. The benefit of PPPs is also accentuated by these two projects, which proved vital in ensuring the successful roll-out and scaling up of online services in the public sector.

Lastly, a key finding from the e-Ghana project worth mentioning is the importance of securing funding for maintenance and operation activities following project completion. The close out report of the e-Ghana project noted that the WAN established during the project was at risk of collapsing at the end of the project due to a lack of planned maintenance.

2.4.2.4 India's Digital Identification Project—Aadhaar

In 2011, the World Bank estimated that 21.2% of India's population survived on less than US$ 1.90 a day.[221] With a population of over 1.35 billion people, this translates to more than 286 million people living in extreme poverty. New data suggests that India has managed to reduce the percentage of its population living in extreme poverty to 5.1%. However, the UN Development Programme's 2019

---

[219]Ibid.

[220]The World Bank, *Ghana Economic Transformation Project*, https://projects.worldbank.org/en/projects-operations/project-detail/P166539?lang=en, accessed: 28 February 2020.

[221]The World Bank, *Poverty & Equity Portal—India*, The World Bank Group, 2020, http://povertydata.worldbank.org/poverty/country/IND, accessed: 17 March 2020.

Multidimensional Poverty Index found 369.55 million people in multidimensional poverty[222] in India in 2016/2017.[223]

With so many people living in poverty in India, it is no surprise that subsidies and social grants account for a large portion of the country's economy. The food, fuel and fertilizer subsidy in India amounted to 1.73 trillion Indian Rupees in 2010[224] (approximately US$ 40 billion in 2019[225]). However, corruption, poor transparency and fake/ghost registrations crippled these support structures. Official estimates indicate that in 2011/2012, around half of the kerosene, wheat and the sugar subsidies did not reach their intended targets.[226]

A large contributor to the inefficient subsidy programmes was the lack of a traceable national identity card that could not be duplicated. In 2008[227] the Government of India created the Unique Identification Authority of India (UIDAI)[228] with the mandate to generate and issue Unique Identification numbers (UID) to all residents of India and ensure the UID eliminated duplicate and fake entities, as well as provide easy and quick verification and authentication.[229]

The solution the UIDAI selected was the Aadhaar. Aadhaar is a Hindi word, which means foundation and refers to a unique twelve digit number that is assigned to a resident of India and linked to their biometric and demographic identity. The number is randomly generated, and no personal data is encoded within the numbers or their sequence. The information associated with the Aadhaar number is shown below:

1. **Demographic Data**: Name, date of birth (verified) or age (declared), gender, address, mobile number, and email address (optional).[230]

---

[222]Multidimensional poverty assesses not only financial shortages; it includes lack of electricity, access to clean water, access to healthcare, education levels and also nutrition levels.

[223]N. McCarthy, *Report: India lifted 271 million people out of poverty in a decade*, Forbes, July 2019, www.forbes.com/sites/niallmccarthy/2019/07/12/report-india-lifted-271-million-people-out-of-poverty-in-a-decade-infographic/#53cc5f442284, accessed: 17 March 2020.

[224]P. Misra, *Lessons from Aadhaar: Analog aspects of digital governance shouldn't be overlooked*, Pathways for Prosperity Commission Background Paper Series: no. 19, Oxford, United Kingdom, 2019.

[225]Inflation Tool, *Inflation calculator—India Rupee*, Inflation Tool, 2020, www.inflationtool.com/indian-rupee?amount=100&year1=2010&year2=2019, accessed: 17 March 2020.

[226]The specific percentages of lost subsidies in 2011/2012 are: 41% of the kerosene subsidy, 54% of the wheat subsidy, and 48% of the sugar subsidy (2014–15 Economic Survey, Government of India).

[227]Originally in 2008, UIDAI was part of the Planning Commission, however the UIDAI was reallocated to the Department of Electronics & Information Technology of the Ministry of Communications & Information Technology in 2015, and in 2016 UIDAI became a statutory authority and government department under the Aadhaar Act 2016 (Unique Identification Authority of India, *Aadhaar Ecosystem)*, https://uidai.gov.in/aadhaar-eco-system.html, accessed: 17 March 2020.

[228]P. Misra, *Lessons from Aadhaar: Analog aspects of digital governance shouldn't be overlooked*, 2019.

[229]Unique Identification Authority of India, *Aadhaar Ecosystem*, Government of India, Jan 2019, https://uidai.gov.in/aadhaar-eco-system.html, accessed: 17 March 2020.

[230]Unique Identification Authority of India, *What is Aadhaar*, Government of India, Jan 2019, https://uidai.gov.in/what-is-aadhaar.html, accessed: 18 March 2020.

2. **Biometric Data**: Ten fingerprints, two iris scans and a photograph of the individual's face.[231]

According to the UIDAI, Aadhaar "is unique and robust enough to eliminate duplicates and fake identities and may be used as a basis/primary identifier to roll out several Government welfare schemes and programmes for effective service delivery thereby promoting transparency and good governance".[232] Furthermore, "it is a strategic policy tool for social and financial inclusion, public sector delivery reforms, managing fiscal budgets, increasing convenience, and promoting hassle-free people-centric governance".[233]

Aadhaar is unique in the following ways:

- The number itself can be verified without the need for a physical card (although cards are issued for ease of use).[234]
- It does not contain any intelligence or personal data encoded within the number.[235]
- An Aadhaar number is proof of identity but does not confer citizenship and, as such, can be issued to any resident of India.[236]
- An individual can register for an Aadhaar number even if they do not have any documentation to prove their demographic data, provided someone vouches for the individual via the "introducer system".[237]

Registration for an Aadhaar number is free of cost and involves providing the necessary biometric and demographic information at an enrolment centre or registrar. The information is encrypted and transferred to the Central Identities Data Repository (CIDR) where it is compared with existing data to search for duplicate information. This process is called deduplication and any duplicate data are verified by CIDR staff and, if necessary, the enrolment centre and individual are contacted. After successful deduplication, the Aadhaar number is generated and assigned to the applicant who is enrolled in the CIDR. An Aadhaar card (a card which summarizes an individual's Aadhaar number and personal data, as shown in Fig. 2.22) is issued to the applicant's address by post. An electronic version of the Aadhaar number (referred to as an e-Aadhaar) is created and made available online for the applicant to download.

Through Aadhaar, an individual's identity and other demographic data can be verified using their Aadhaar number at any given time or place using the digital online platform named Aadhaar Authentication, which has various authentication

---

[231] Ibid.

[232] Ibid.

[233] Ibid.

[234] P. Misra, *Lessons from Aadhaar: Analog aspects of digital governance shouldn't be overlooked*, 2019.

[235] Unique Identification Authority of India, *What is Aadhaar*, Government of India, Jan 2019, https://uidai.gov.in/what-is-aadhaar.html, accessed: 18 March 2020.

[236] Ibid.

[237] P. Misra, *Lessons from Aadhaar: Analog aspects of digital governance shouldn't be overlooked*, 2019.

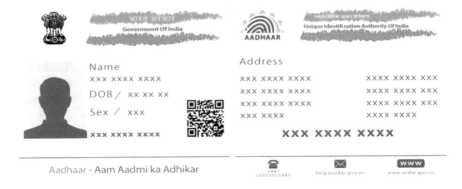

Fig. 2.22 Example of an Aadhaar Card (front on left and back on right) (BankBazaar, *Aadhaar Card*, BankBazaar, 2020, www.bankbazaar.com/aadhar-card.html, accessed: 18 March 2020.)

modes described below. The authentication process compares information stored in the UIDAI's CIDR and not on the Aadhaar card.[238]

1. **Demographic Authentication**: This authentication mode verifies the supplied Aadhaar number and demographic data with the data contained in the CIDR.
2. **One-Time Pin Based Authentication**: This mode issues a one-time pin (OTP), valid for a limited time period, to the mobile number or email address of an Aadhaar user. The OTP is then submitted and verified to a service delivery provider along with the Aadhaar number of the user.
3. **Biometric Authentication**: This authentication mode allows service delivery providers to validate end-users using fingerprint or iris scanners.
4. **Multi-factor Authentication**: This two-factor authentication mode uses a combination of the above authentication modes.

The different authentication modes return two different types of results, depending on the application. The first result is a simple "Yes or No" response, which verifies if the supplied data is the same as the data contained within the CIDR for a specific Aadhaar number. The second type of result is referred to as an "e-KYC" authentication (and refers to an electronic know-your-customer authentication) and is often used in financial applications. These authentications return proof of identity of the individual, their physical address, gender and date of birth.[239],[240] The communication

---

[238]Unique Identification Authority of India, *Aadhaar Ecosystem—Authentication Ecosystem*, Government of India, Jan 2019, https://uidai.gov.in/aadhaar-eco-system/authentication-ecosystem.html, accessed: 25 March 2020.

[239]Unique Identification Authority of India, *Aadhaar Ecosystem—Authentication Ecosystem*, Government of India, Jan 2019, https://uidai.gov.in/aadhaar-eco-system/authentication-ecosystem.html, accessed: 25 March 2020.

[240]P. Misra, *Lessons from Aadhaar: Analog aspects of digital governance shouldn't be overlooked*, 2019.

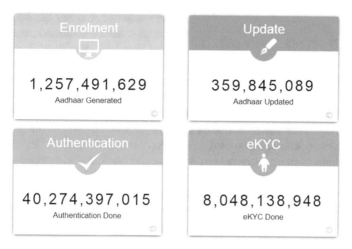

Fig. 2.23 Aadhaar Dashboard data as of April 2020 (Unique Identification Authority of India, *Welcome to AADHAAR Dashboard*, Government of India, April 2020 (data updated monthly), https://uidai.gov.in/aadhaar_dashboard/, accessed: 1 April 2020)

occurs over the Internet using either mobile or fixed line networks.[241] To overcome connectivity issues, Smart QR Codes are issued with e-Aadhaars and Aadhaar letters, which can be read by an application developed and released by the UIDAI, thereby validating an individual without the need for an active Internet connection. The QR code contains a photograph of the Aadhaar holder along with their demographic information (name, gender, date of birth, mobile number, email address and physical address) and also contains a 2048-bit digital signature for added security.[242]

India has enrolled over 1.25 billion people using the Aadhaar system, achieving a coverage rate of 99% of its adult population.[243] These numbers indicate that the Aadhaar project was successful in providing residents of India with unique identifications. The Aadhaar project has also been successful in promoting financial inclusion and optimizing subsidies and public service delivery in the country. There have been over 40.2 billion "Yes or No" Aadhaar authentications and over 8 billion e-KYC checks, as shown in Fig. 2.23.[244]

---

[241] Unique Identification Authority of India, *Aadhaar Ecosystem—Authentication Ecosystem*, Government of India, Jan 2019, https://uidai.gov.in/aadhaar-eco-system/authentication-ecosystem.html, accessed: 25 March 2020.

[242] Unique Identification Authority of India, *Secure QR Code Reader*, Government of India, Jan 2019, https://uidai.gov.in/ecosystem/authentication-devices-documents/qr-code-reader.html, accessed: 1 April 2020.

[243] P. Misra, *Lessons from Aadhaar: Analog aspects of digital governance shouldn't be overlooked*, 2019.

[244] Unique Identification Authority of India, *Welcome to AADHAAR Dashboard*, Government of India, April 2020 (data updated monthly), https://uidai.gov.in/aadhaar_dashboard/, accessed: 1 April 2020.

From a financial inclusion perspective, the impact of Aadhaar has been vast. As of April 2017, more than 260 million bank accounts had been opened using Aadhaar as proof of identity. Prior to Aadhaar, formal proof of identification was one of the major reasons for the lack of formal bank accounts. In 2010 almost two thirds of the population did not have bank accounts. Aadhaar, coupled with the Pradhan Mantri Jan Dhan Yojana (PMJDY) programme launched in 2014[245] has reduced the number of Indians without bank accounts to less than 20%.[246] The PMJDY financial inclusion programme was launched by the Indian Prime Minister Narendra Modi and enables bank accounts to be opened with no initial capital requirement. It resulted in over 18 million bank accounts being opened in one week, a world record recognized by the Guinness World Records.[247] According to the PMJDY website, more than 382 million people have set up bank accounts using the scheme.

The National Payments Corporation of India (NPCI[248]) utilized Aadhaar to establish the Aadhaar Payment Bridge (APB) System, which enables the Government of India to pay social grants and subsidies directly to beneficiaries' bank accounts[249] under the Direct Benefit Transfer (DBT) Scheme, employing Aadhaar numbers as the unique identifier. The APB System also has the functionality to assist with settlement of people and supports rent seekers in finding potential accommodation.[250] The Government of India stated that at the end of 2017, 394 government schemes were making use of the APB System and estimated that approximately US\$ 37.38 billion had been dispersed through APB.[251] Aadhaar, along with the DBT scheme, is assisting with remittances for the Mahatma Gandhi National Rural Employment Guarantee Act (MGNREGA), which aims to reduce poverty in rural

---

[245]Department of Financial Services, *Pradhan Mantri Jan Dhan Yojana (PMJDY)*, Ministry of Finance, Government of India, https://pmjdy.gov.in/about, accessed: 22 April 2020.

[246]P. Misra, *Lessons from Aadhaar: Analog aspects of digital governance shouldn't be overlooked*, 2019.

[247]Kevin Lynch, *India makes financial world record as millions open new bank accounts*, Guinness World Records, January 2015, www.guinnessworldrecords.com/news/2015/1/india-makes-financ ial-world-record-as-millions-open-new-bank-accounts, accessed: 22 April 2020.

[248]The NPCI was established with guidance from the Reserve Bank of India and the Indian Banks Association and is the organizations responsible for operating retail payments and settlement systems in India (National Payments Corporation of India, *About Us*, www.npci.org.in/about-us-background, accessed: 1 April 2020).

[249]The software solution, which enables the integration of the various systems is open source and is named IndiaStack, which comprises a set of standard application programme interfaces with four specific layers—the "presenceless layer" which utilizes a biometric digital identity, the "paperless layer" which links digital records to a digital identity, a "cashless layer" which comprises a single interface to all of India's bank accounts and a "consent layer" which enables free and secure transmission of data (IndiaStack, www.indiastack.org/), accessed: 4 April 2020.

[250]National Payments Corporation of India, *Frequently asked questions on Aadhaar Payment Bridge System*, National Payment Corporation of India, Mumbai, www.ucobank.com/pdf/faq-apb. pdf, accessed: 1 April 2020.

[251]P. Misra, *Lessons from Aadhaar: Analog aspects of digital governance shouldn't be overlooked*, 2019.

areas by providing a minimum of 100 days of wages to households whose adult members perform unskilled manual work.[252]

Aadhaar numbers have been linked to numerous benefit schemes across India, including liquefied petroleum gas (LPG) subsidies, welfare housing programmes, house allotments, education funding, driving licenses, insurance policies, loan waivers and pension funds.[253] Prior to Aadhaar, pensioners were required to provide a physical life certificate to their bank in person every year in order to secure their pension for the following year. The Jeevan Pramaan Project revolutionized this process by creating digital life certificates linked to a pensioner's Aadhaar number.[254] Ensuring the continuity of an individual's pension could now be done online. This resulted in an increase in the number of people enrolled in the country's pension fund from 1.65 million in 2016 to over 15 million at the end of 2017.[255] This is an example of how Aadhaar has transformed public service delivery in India from ineffective paper-based systems to efficient, paperless transactions with increased inclusivity.

Another interesting application of Aadhaar is an online digital locker assigned to each Aadhaar number that allows users to store important documents such as health records, academic records, driving licenses and other documents, online. The digital locker has digital signing functionality and allows for certain documents to be automatically verified and recognized by certain government departments.

The Indian government is also investigating linking Aadhaar numbers to motor vehicles in order to implement an automatic tolling system and is also exploring the possibility of using Aadhaar to realize an online e-voting system.[256]

Other successes include a tangible reduction in fraudulent transactions and corruption. The Government of India reported over 30 million fake LPG connections were identified through Aadhaar, along with 23 million fake food ration cards.[257] Additionally, in 2018, the UIDAI claimed that as of 31 March 2018, Aadhaar had resulted

---

[252]R. Raju et al., *Aadhaar Card: Challenges and Impact on Digital Transformation*, Manav Rachna International University, Faridabad, India.

[253]R. Raju et al., *Aadhaar Card: Challenges and Impact on Digital Transformation*, Manav Rachna International University, Faridabad, India.

[254]The Economic Times, *Everything a pensioner needs to know about submitting Jeevan Pramaan Patra online*, The Economic Times, India Times, Nov 2019, https://economictimes.indiatimes.com/wealth/save/everything-a-pensioner-needs-to-know-about-submitting-jeevan-pramaan-patra-online/articleshow/72196725.cms?from=mdr, accessed: 4 April 2020.

[255]P. Misra, *Lessons from Aadhaar: Analog aspects of digital governance shouldn't be overlooked*, 2019.

[256]R. Raju et al., *Aadhaar Card: Challenges and Impact on Digital Transformation*, Manav Rachna International University, Faridabad, India.

[257]P. Misra, *Lessons from Aadhaar: Analog aspects of digital governance shouldn't be overlooked*, 2019.

Table 2.6 Assessing financial inclusion gaps in India between 2011 and 2017

Demographic	Percentage with bank account		
	2011 (%)	2017 (%)	World average (2017) (%)
Overall population	35	80	67
Gender			
Men	44	83	71
Women	26	77	64
Employment status			
Employed	44	84	72
Unemployed	26	75	58
Education level			
Secondary education and higher	59	85	77
Primary education or less	31	75	54
Income Level			
Richest 60%	41	82	72
Poorest 40%	27	77	59

A. Demirguc-Kunt et al., *The Global Findex Database 2017: Measuring financial inclusion and the fintech revolution*, The World Bank Group, 2018

in government savings of more than US$ 23.9 billion.[258] Concerns have, however, been raised regarding the method and data used to arrive at this figure.[259]

Aadhaar and Aadhaar-enabled payment schemes have successfully promoted financial inclusion and reduced gaps across gender, income, and education levels, as highlighted below in Table 2.6. Included in the table are the world averages for each demographic. It is clear from the data that India exhibits higher financial inclusion than the world averages across every demographic.

The OECD 2018 Global Trends Report on Embracing Innovation in Government states that "Aadhaar is the largest identity programme ever created and the potential for its use as a platform is unprecedented".[260] The report elucidates how

---

[258]The Economic Times, *Aadhaar-enabled DBT savings estimated over Rs 90,000 crore*, The Economic Times, India Times, July 2018, https://economictimes.indiatimes.com/news/economy/finance/aadhaar-enabled-dbt-savings-estimated-over-rs-90000-crore/articleshow/64949101.cms, accessed: 3 April 2020.

[259]P. Misra, *Lessons from Aadhaar: Analog aspects of digital governance shouldn't be overlooked*, 2019.

[260]OECD, *Embracing Innovation in Government: Global Trends 2018*, OECD Publishing, Paris, 2018, www.oecd.org/innovation/innovative-government/embracing-innovation-in-government-2018.pdf, accessed: 3 April 2020.

Aadhaar's successes are broad-reaching and how citizens are receiving better service delivery wait times for Public Distribution System[261] (PDS) food supplements, which have been reduced from hours to minutes. Increased information and mobility allow customers to make informed decisions about which shops to receive their rations from. Using the same data, the government can monitor the performance of shops and uphold the quality of services. This increases the people's trust of the government, as well as community wellbeing as people know they cannot be cheated out of their benefits.[262]

While the restructuring of public services as a consequence of Aadhaar has been profoundly widespread and generally successful, Aadhaar has not been without failures. It has been marred by numerous issues surrounding data security and data privacy. In early 2018 it was reported that one billion Aadhaar numbers had been jeopardized and could be purchased via WhatsApp for 500 Indian Rupees, along with the associated demographic data.[263] The government has, however, maintained that there has not been a data breach of the UIDAI's CIDR. It has also argued that if an individual's Aadhaar data is jeopardized, it does not pose a large threat since Aadhaar requires additional authentication through biometric or OTP methods.[264]

Issues relating to data privacy have also tainted the Aadhaar system, with certain citizens being concerned that the Indian government could possibly collect and misuse their personal data as a result of Aadhaar. These sentiments were exacerbated by the lack of data protection and data privacy legislation in India at the time Aadhaar was established and for many years thereafter. It was only in December 2019 that the Personal Data Protection Bill was tabled in parliament. The bill offers a framework on how a person's data may be processed and mandates the establishment of a Data Protection Authority. Yet, it is worth noting that the bill has specific exemptions for data privacy relating to national security issues, which has raised anxieties both locally and internationally.[265] In particular, government institutions can gain full access to an individual's personal data where they suspect terrorist or associated intentions.

Aadhaar has become a case study that demonstrates the importance of adequate legal frameworks for data protection, security and access, especially for countries

---

[261] The Public Food Distribution system is a welfare scheme which distributes food supplements (such as rice and grains) to poor people in India and was established in 1965 (R Raju et al., *Aadhaar Card: Challenges and Impact on Digital Transformation*, Manav Rachna International University, Faridabad, India).

[262] Ibid.

[263] S. Vijayasarathy, *Your Aadhaar number on sale for Rs 500, all Aadhaar-linked details of 1 billion Indians leaked*, India Today, New Delhi, January 2018, www.indiatoday.in/techno logy/news/story/aadhaar-number-on-sale-for-rs-500-linked-details-of-1-billion-indians-leaked-1122086-2018-01-04, accessed: 4 April 2020.

[264] P. Misra, *Lessons from Aadhaar: Analog aspects of digital governance shouldn't be overlooked*, 2019.

[265] H. Arakalai, *The Personal Data Protection Bill could be a serious threat to Indians' privacy*, Forbes India, December 2019, www.forbesindia.com/article/leaderboard/the-personal-data-protec tion-bill-could-be-a-serious-threat-to-indians-privacy/56623/1, accessed: 4 April 2020.

pursuing digital identification solutions. Such legislation should cover both user consent and notification and also ensure that an independent authority is established to oversee enforcement of data usage and regulations.[266]

With Aadhaar linked to PDS, which many people rely on for their survival, it is of paramount importance that the authorization system is reliable and always accessible. Troublingly, the IDInsight report in 2018 revealed a PDS exclusion rate of between 0.8 and 2.2% due to Aadhaar-related failures across three Indian states. This translated to over two million people being denied their food rations.[267] Such issues have resulted in people questioning the constitutionality of the government making it compulsory for citizens to link their Aadhaar data in order to receive social benefits and subsidies.[268]

2.5 Conclusion

This chapter assessed global e-governance trends and current e-governance levels in Africa through a detailed analysis of the UN *E-Government Survey* report from 2018. The analysis revealed a positive trend in e-government adoption and readiness levels globally and noted that all 193 UN member states exhibit some form of online presence. Africa exhibited a similar positive trend, albeit at a significantly slower rate. The 2018 Survey revealed that 13 African countries moved from the "Low" e-governance classification to "Middle". However, in comparison to global performances in the e-government assessment, Africa, as a region, underperformed considerably, where only four African countries achieved e-governance scores higher than the world average.

It was demonstrated that the provision of a unique identification document for all citizens is a fundamental governance issue that must be addressed before e-governance systems can be implemented. Without verifiable identification, people cannot open bank accounts, register for schooling, apply for loans, start a business or receive government grants and subsidies. Therefore, it is critical that innovative identification systems are established as a matter of urgency throughout Africa. Examples demonstrating the large scale impact of successful identification systems on financial inclusion and social welfare schemes were provided.

Both globally and within Africa, increased e-governance is translating into increased inclusivity of minority demographics. 57% of African countries offer targeted online services for vulnerable groups. Moreover, e-government has amplified transparency, accountability and openness in governance. The majority of UN

---

[266] J. Attick, *Digital identity: The essential guide*, ID4Africa Identity Forum, 2014, www.id4africa. com/main/files/Digital_Identity_The_Essential_Guide.pdf.

[267] S. Sen, *A decade of Aadhaar: Lessons in implementing a foundational ID system*, ORF Issue Brief No. 292, May 2019, Observer Research Foundation.

[268] P. Misra, *Lessons from Aadhaar: Analog aspects of digital governance shouldn't be overlooked*, 2019.

member states have online e-procurement platforms, online announcements of public tenders and online public procurement evaluation data. The number of open government data portals publishing government datasets pertaining to education, health, environmental, social programmes and finance departments increased in 2018. The promotion of citizen participation in policy, legislation and service design and delivery through e-governance was discussed.

The 2018 *E-Government Survey* established that public-private partnerships are the most successful recent model for e-government implementation due to the scale and interdisciplinary nature of e-governance projects. Accordingly, Africa has exhibited the highest regional increase, with 44 African countries establishing online services through partnerships with the private sector. Such partnerships enable shared risk profiles, increased resources and vertical integration of services but require detailed contracts complete with risk assessments and performance indicators to safeguard all involved parties. Thus, to promote e-governance, governments must provide supportive economic environments that facilitate public-private partnerships and multi-stakeholder partnerships. The majority of innovative e-governance examples examined in this document were implemented using public-private partnerships.

Africa's comparatively poor levels of e-government and e-governance readiness are attributed to its underdeveloped ICT infrastructure (which is the least developed in the world) and the resultant low Internet connectivity rates. While mobile Internet access provides an effective solution for reaching rural areas with little fixed-line infrastructure, the relatively high cost is currently prohibitive in many areas. The UN's 2018 survey and the ITU's GCI report both confirm that Internet access remains a significant barrier for e-governance in Africa. Affordable, high-bandwidth Internet access is a basic requirement for e-governance. In less developed countries with poor ICT infrastructure, e-government strategies need to include both digital and non-digital (such as call centres, post offices, television, radio and face-to-face services) channels to guarantee service provision to the entire population. In developing countries, well-defined, national broadband and ICT development strategies are paramount for addressing the connectivity divide, and it was demonstrated how governments can leverage the myriad benefits offered by mobile technologies to assist with governance activities.

Digital literacy issues must also be addressed when implementing e-governance in order to ensure inclusivity. This can be especially difficult in certain African countries with poor literacy rates. It was shown that governments can overcome these issues by providing face-to-face services at Internet-connected kiosks that are distributed strategically around a country.

National strategies and policies that specifically address e-government and ICT development must be established as a prerequisite for e-governance. Data privacy and security legislation is also critical to ensure the safe and ethical use of citizens' data. Similarly, national technical standards and guidelines must be developed and adopted to facilitate the interoperability of systems. An open source data exchange layer protocol was presented that provides data transmission amid governmental departments with maximum interoperability and avoids system architecture overhauls and hardware replacement wherever possible.

As governments increase their e-governance services, the risk and severity of cyberattacks rises. Moreover, the interdependencies of online public platforms can result in potentially catastrophic disruptions in the health, safety, security and social sectors from such attacks. Without resilient cybersecurity measures, governments expose both themselves and their citizens to these risks. Therefore, it is crucial that governments establish comprehensive and responsive cybersecurity legislation, overseen by a dedicated government institution.

Other vital aspects of cybersecurity that must be adequately addressed by governments implementing e-government applications were shown to include: establishing national Computer Emergency Response Teams (CERTs), implementing National Cybersecurity Strategies, and ensuring sufficient capacity within the sector by running public awareness campaigns and providing certification and accreditation for professionals in the field. It was established that African states must increase their cybersecurity endeavours to ensure the resilience of their e-government systems.

Academics and governments are researching the application of new technologies such as big data analytics, artificial intelligence, machine learning and blockchain to develop new e-governance solutions. Integrating and combining these new technologies facilitates new e-governance models and architectures. Governments should promote research and development in new technologies that have applications in e-governance.

Blockchain offers new possibilities for realizing e-voting and online voting schemes due to its decentralized nature and immutable distributed ledgers. It was demonstrated how blockchain can be employed to overcome many of the hurdles facing e-voting and electronic voting machines. Voting solutions employing blockchain technology could provide African countries with decentralized, publicly verifiable elections, which will promote democracy throughout the continent.

Chapter 3
Space Supporting Governance Through Connectivity and Communication

Abstract This chapter focuses on the connectivity required to achieve good governance, both in terms of e-governance through virtual voting and other mechanisms, and through the other capabilities that are enabled through better connectivity, such as education and awareness of issues that directly and indirectly relate to governance. An overview of current space-based Internet and communication technologies is provided as well as a thorough overview of the constellation-based Internet technologies that will be available in the near-future, with widespread possible benefits for governance in Africa.

3.1 Introduction

High speed broadband Internet to locations where access has been unreliable, expensive, or completely unavailable.[1]

Elon Musk (Starlink)

Good governance requires the ability to monitor a region's resources, both natural and human-made. Space technologies enable overwatch of almost everything that may support a country or region and facilitate faster and more accurate governance decision making.

This chapter examines the specific way in which governance can be improved through the use of space technologies. It should be noted that these technologies must be used in conjunction with systems on the ground. Space technologies mainly allow communication and visualisation and cannot themselves perform any of the actions necessary to improve governance. While these technologies do not necessarily directly contribute to the ideas and projects mentioned in Chaps. 1 and 2, they make vast indirect contributions, forming a backbone for all digital governance activities.

---

[1]TIMESOFINDIA.COM, Elon Musk's 'Internet from space' service coming in six months, 24 April 2020, https://timesofindia.indiatimes.com/gadgets-news/elon-musks-Internet-from-space-service-coming-in-six-months/articleshow/75356989.cms (accessed 20 Mai 2020).

© The Editor(s) (if applicable) and The Author(s), under exclusive license to Springer Nature Switzerland AG 2020
A. Froehlich et al., *Space Supporting Africa*, Studies in Space Policy 28,
https://doi.org/10.1007/978-3-030-52260-5_3

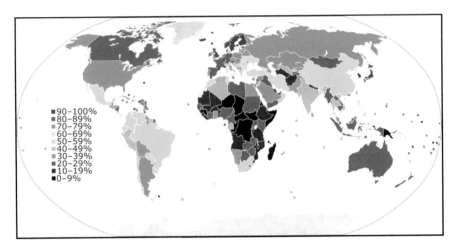

Fig. 3.1 Percentage of Internet users per country worldwide (Jeff Ogden (W163) [CC BY-SA 3.0 (https://creativecommons.org/licenses/by-sa/3.0)], "Percentage of Internet users worldwide", 2016, https://commons.wikimedia.org/wiki/File:InternetPenetrationWorldMap.svg (accessed 26 November 2019))

Currently, Africa suffers the highest lack of connectivity of any continent, with even Antarctica having higher average connection speeds per person present. Only about 20% of Africans have some form of Internet connection, much of which has extremely poor performance in terms of bandwidth, data allocation and latency.[2]

Because of Africa's large surface area and harsh terrain, very little cable-based connectivity exists in the region, and with huge upfront costs, enterprises are unwilling to invest in infrastructures necessary to establish fibre or copper-based Internet connectivity. Copper is especially an issue as cable theft is highly prevalent throughout the African continent.[3] Figure 3.1 shows just how far behind Africa is in terms of connectivity.

Satellite-based Internet has existed commercially since 2003,[4] with its extraterrestrial position enabling Internet connection to any location on Earth covered by the satellite's footprint, making it advantageous for use in Africa.[5]

---

[2]"ICT Facts and Figs. 2005, 2010, 2017", *Telecommunication Development Bureau, International Telecommunication Union (ITU)*.

[3]Mitch Mitchell, "The real cost of electrical cable theft to the SA economy: R187 billion per year", 2019, https://www.pressportal.co.za/energy-and-environment/story/17875/the-real-cost-of-electrical-cable-theft-to-the-economy-r187-billion-per-year.html (accessed 26 November 2019).

[4]"First Internet Ready Satellite Launched", 29 September 2004, Space Daily.

[5]See, for example: C. Kotze, *A Broadband Apparatus for Underserves Remote Communities: Connecting the Unconnected* (Cham: Springer, 2020); T. Hugbo, "The Importance of Internet Accessibility and Smart City in Sub-Sahara African Region Through Space Technology", in *Embedding Space in African Society: The United Nations Sustainable Development Goals 2030 Supported by Space Applications*, ed. A. Froehlich (Cham: Springer, 2019), 105–112; and A.S. Martin, "Internet by Satellite for Connecting the African Continent: A Glance on the Partnership Between Rwanda

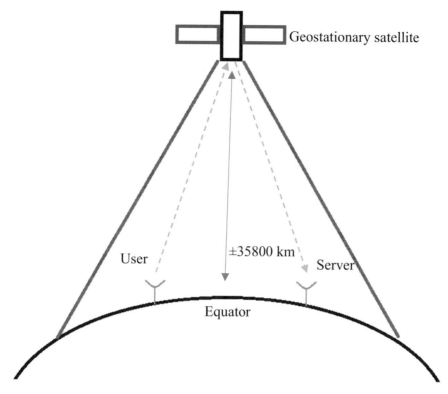

Fig. 3.2 Geostationary configuration

This chapter will look at existing and future space-based communication technology, as well as criticisms thereof, and potential benefits for governance and Africa.

3.2 Current Geostationary Internet Technologies

There are currently a number of competing geostationary Internet providers. A Geostationary Orbit (GEO) satellite orbits the equator at a distance of about 35,800 km from the Earth's surface. This means that it orbits at the same speed that Earth rotates and therefore always covers the same region of Earth. Figure 3.2 shows this configuration.

and the Private Company OneWeb", in Space Fostering African Societies: Developing the African Continent through Space, Part 1, ed. A. Froehlich (Cham: Springer, 2020), 61–70.

3.2.1 Technical Attributes

The principle benefits of a geostationary Internet satellite are that it can cover a large area and is always available, subject only to the weather.

However, because of the huge distance, there are several drawbacks. The first is latency (or ping), defined in milliseconds where lower is better. Latency is different to Internet speed in that it defines the response time of a server to a user's request. In simple terms it may be thought of as the time needed to find a website, while Internet speed would be the time needed to load the said website. In geostationary Internet, the speed of light heavily limits latency. Equation 3.1 shows the physical time needed for a signal to travel to or from a geostationary satellite. Speed is the speed of light in m/s.

$$
\begin{aligned}
Latency\ to\ satelltie &= \frac{Distance}{Speed} \\
&= \frac{358000000}{3 \times 10^8} \\
&= 119.3\ \mathrm{ms}
\end{aligned}
\tag{3.1}
$$

$$
\begin{aligned}
Total\ latency &= latency\ to\ satellite \times 4 \\
&= 119.3 \times 4 \\
&= 477.2\ \mathrm{ms}
\end{aligned}
\tag{3.2}
$$

Equation 3.2 shows the total minimum theoretical latency for a request from a user to a satellite, and then the receipt of the request from a server, hence multiplying by four. Since most users and servers will be further than 35,800 km apart, as they will not be directly underneath the satellite, and some time is needed for processing, the average latency for geostationary satellite Internet is around 600 ms. This makes video calling or other real-time applications extremely challenging, as there will be almost a full second of delay between requests and responses.

The second disadvantage is that the signal is of very low intensity after travelling such a distance. While this originally limited the speed, this has been improved through the use of the Ka-band (up to 40 GHz) to be able to reach gigabit speeds. However, because the signal is small, it handles interference poorly and requires a large gain on either side. This means that inclement weather can affect the signal heavily (known as rain fade), and that large receivers are necessary, often needing small dishes (known as very small aperture terminals—VSAT) to maintain good connectivity (Fig. 3.4). The satellites themselves have to be very large (solar arrays can be 50 m or longer), as seen in Fig. 3.3, and consume large amounts of power.

Lastly, the geostationary orbit has a limited amount of space. Because of this, there is a limit to the possible number of Internet satellites that can be put there.

Fig. 3.3 ABS 2A and Eutelsat 115 W stacked for launch. Photo from SpaceX

Fig. 3.4 Example of a deployable VSAT dish (Bidgee [CC BY 3.0 (https://creativecommons.org/lic enses/by/3.0)], "Very Small Aperture Terminal", 2006, https://commons.wikimedia.org/wiki/File: Bigpond_Internet_Satellite.jpg (accessed 26 November 2019))

Also, because of the distance from Earth, it is virtually impossible to de-orbit from this location (due to both lack of atmospheric drag and the amount of fuel it would need), and satellites are placed further out into what is known as a graveyard orbit

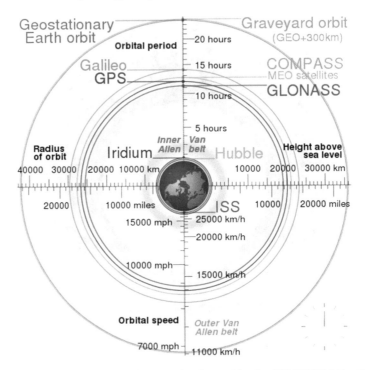

Fig. 3.5 Comparison of different orbital operating figures (Cmglee [CC BY 3.0 (https://creativec ommons.org/licenses/by/3.0), "Orbital Diagram", 2019, https://commons.wikimedia.org/wiki/File: Comparison_satellite_navigation_orbits.svg (accessed 26 November 2019))

when they have completed their active lifespan. It is also very expensive to reach this orbit, incurring high launch costs for geostationary operators. Figure 3.5 shows the comparative distance of geostationary orbits to those much lower, as well as the graveyard orbit.

3.2.2   Cost and Availability of Geostationary-Based Internet in Africa

Since geostationary satellites cover one specific area of the world and cannot move in relation to the Earth, most Internet satellites are currently located over North America, Europe and Asia. However, there are some that provide access to Africa.

Al Yah Satellite Communications (Yahsat) launched the Y1B satellite in 2012, which uses multi-spot Ka-band to deliver Internet to most African regions.[6] Multi-spot uses multiple feed antennas to reflect off the same large dish at multiple frequencies, reducing interference and allowing a much greater multiple of users, the most recent advancement in geostationary Internet provision. Yahsat offers up to 7 mbps downlink and 1.5 mbps uplink. However, this costs US$ 1038 per month as of 2019, and has a limit of 100 gb,[7] which is far more expensive than land or cellular-based services.

Per-country options may be slightly cheaper. Vodacom, a telecoms company in South Africa, advertises 4 mbps satellite Internet with a maximum data usage of 30 gb for US$ 150 per month. It does not disclose which satellite it uses.[8] However, South Africa has large amounts of infrastructure compared to most African countries, and it has to be priced lower to remain competitive.

3.3 Large-Constellation Satellite Internet Background

In order to address the disadvantages mentioned earlier, and the high price-points, large constellation based Internet is being developed. Large constellation (or mega-constellation) means a massive number of satellites operating together to serve one purpose or network. These satellites are usually much smaller than the single large satellites that currently provide access to the Internet. Figure 3.6 shows a simulation of SpaceX's Starlink network, providing a theoretical latency of 82 ms between London and Johannesburg, far faster than that of the current Internet.

There are two primary reasons why these constellations can be developed. The first is the overall improvement of computer processing. Processors are now significantly smaller, more powerful and more energy-efficient than they were in the early years of the space age. A modern smartphone is powerful enough to control the launch of 1 million Saturn V's (12,190 instructions per second vs 16 billion (2 GHz processor and 8 cores). This is in accordance with Moore's law,[9] which states that every two years the number of transistors on a piece of silicon doubles. Figure 3.7 shows the real-life proof of this trend, showing the decrease in the size of flash memory.

This allows very small satellites to have ample processing power to perform even the most complex of computing tasks, such as the signal processing needed for large-scale communications. Even though processor designers are reaching a physical limit

---

[6]"Yahsat 1B", 2012, https://www.satbeams.com/satellites?norad=38245, (accessed 26 November 2019).

[7]"Yahclick Broadband", https://ts2.space/en/yahclick/ (accessed 1 December 2019).

[8]"Vodacom Satellite Internet", https://www.vodacombusiness.co.za/business/solutions/Internet/bro adband-connect-satellite, (accessed 1 December 2019).

[9]Moore, Gordon E., "Cramming more components onto integrated circuits", *Electronics*, 19 April 1965.

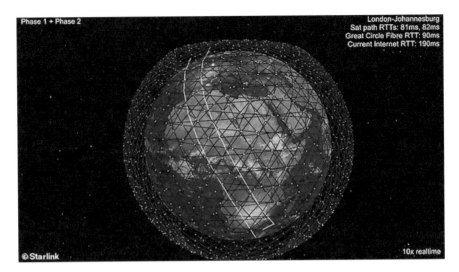

Fig. 3.6 Starlink network simulation (Mark Hadley, *University College London,* 2019)

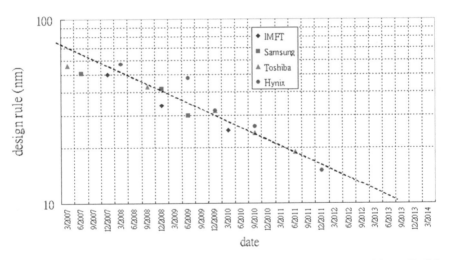

Fig. 3.7 Size of MOSFET scaling of NAND Flash over 7 years (Guiding light at English Wikipedia [CC BY-SA 3.0 (https://creativecommons.org/licenses/by-sa/3.0)], "Decrease in NAND Size", 2011, https://commons.wikimedia.org/wiki/File:NAND_scaling_timeline.png (accessed 1 December 2019))

in terms of transistor size due to electron spacing (the transistors may be too close together and electrons may randomly "jump" between them), quantum computing will enable a new avenue of higher performance to be explored.

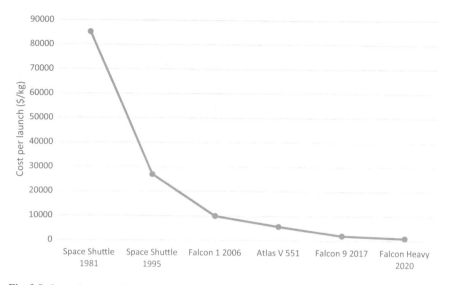

Fig. 3.8 Launch costs of US providers to low Eartb orbit (LEO)

The second is that the cost-of-launch has decreased greatly as well. Rocket Labs provides a dedicated launch for US$ 5.7 million,[10] which is extremely affordable in terms of space. Traditionally, space activities have been dominated by the big governmental actors who could afford the vast amount of time and resources needed to achieve success. Presently, even launch, one of the hardest aspects of space, has become highly commercial. The Russian Soyuz rocket family has changed from a Cold War icon into a transportation method that almost any entity can purchase. SpaceX has been pushing lower and lower costs by focussing on reusability, getting much closer to the unobtained goals of the Space Shuttle than the Space Shuttle ever could. Figure 3.8 shows this decreasing cost of U.S. launch providers.

For these two main reasons, along with other technological improvements such as additive manufacturing and better solar panels, large constellations of satellites that all form a single mega-system can now be launched. The Earth-Observation (EO) company Planet has the world's first working large constellation, operating at least 140 satellites at any point in time. However, EO is in some measures cheaper and less complicated to achieve than communications.

SpaceX's Starlink Internet division[11] is currently leading the mega-constellation Internet gambit. As of 2019 they had launched 120 satellites[12] each weighing 227 kg. Starlink has targeted an operational orbit of 550 km (LEO). This orbit provides two benefits, the first is that latency is extremely low at 25–35 ms, comparable to high-end cellular networks. The second is that this orbit will allow easier de-orbiting

---

[10]Guy Gugliotta, *Air & Space*, 2019, https://www.airspacemag.com/as-next/milestone-180968351/ (accessed 1 December 2019).

[11]"Starlink Website", https://www.starlink.com/.

[12]"Starlink Press Kit", *SpaceX*, 15 May 2019.

Fig. 3.9 Starlink satellites attached to bus. Photo from SpaceX

and eventual natural orbital decay due to atmospheric resistance, which will help minimise space debris. Figure 3.9 shows the second batch of 60 satellites about to be deployed from the rocket bus.

Starlink's satellites incorporate very high-end technology, including phased-array Ku-band antennas and inter-satellite optical links. Starlink has promised speeds of up to 1 Gbps with a full network.[13] SpaceX has further said it will achieve some commercial operational regional coverage by 2020. It will need 480 satellites to achieve minor global coverage.[14] The biggest ambition of SpaceX is to establish a full global constellation of 12,000 satellites. Even with cheaper satellites and much lower cost launch, this will require a huge amount of capital to achieve.

OneWeb is currently the second biggest LEO Internet operator, with six pilot satellites operational and approximately 2000 more planned. OneWeb's satellites are planned for an orbital altitude of about 1200 km,[15] which will make them significantly harder to de-orbit than SpaceX's.

Amazon's Project Kuiper[16] has also planned some 3000 satellites but as of 2019 has not shown or launched any spacecraft.

---

[13] Don Reisinger, "Update on Starlink", *Fortune,* 22 October 2019, https://fortune.com/2019/10/22/elon-musk-twitter-spacex-starlink/ (accessed 14 December 2019).

[14] Elon Musk, "Much will likely go wrong on 1st mission. Also, 6 more launches of 60 sats needed for minor coverage, 12 for moderate" *(Tweet)—viaTwitter,* 11 May 2019.

[15] "OneWeb Satellite Startup to Set up Manufacturing in Florida", *Wall Street Journal,* 3 January 2016.

[16] "Project Kuiper", https://www.amazon.jobs/en/teams/projectkuiper (accessed 14 December 2019).

Fig. 3.10 Satellite train visible from CTIO

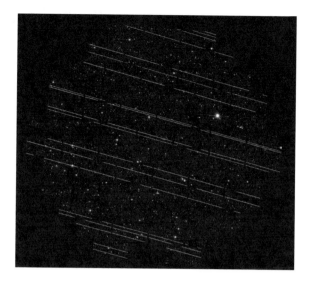

3.3.1 Criticisms of Large Constellation Satellite Internet

A major issue that was identified in 2019 was a by-product of the launch of the Starlink satellites, as illustrated in Fig. 3.10. The large number of satellites at low altitude were extremely visible to astronomers, disrupting many images and measurements, and appearing as a bright train across the night sky.

This is not as serious as it may first appear however, as the satellites will spread out and become orders of magnitude fainter as they reach operational orbit, limiting their disruption to astronomy. However, there is a large difference between 120 and 12,000 satellites, and it remains to be seen what the overall effect will be. SpaceX has endeavoured to paint the next satellites it launches matte black, to further reduce their visibility.

Furthermore, as previously noted, there is increasing worry about a space debris incident, especially when these constellations effectively double the amount of satellites in orbit in the next decade. SpaceX has at least incorporated hall-effect thrusters into its satellites that will enable a degree of collision avoidance and controlled de-orbits. It has even planned to introduce automatic avoidance similar to Tesla's Autopilot software.[17] OneWeb has also declared they are committed to avoiding collisions.

The final major criticism is that these mega-constellations will interfere with existing space-based communications in geostationary orbits. Especially vulnerable

---

[17]Tim Fernholz, "SpaceX's new satellites will dodge collisions autonomously (and they'd better)", *QZ*, 2019, https://qz.com/1627570/how-autonomous-are-spacexs-starlink-satellites/ (accessed 14 December 2019).

will be the existing VSAT dishes, which already have a high sensitivity and low-interference tolerance. Both Starlink[18] and OneWeb[19] have said they will orientate their satellites differently and lower their power level as they pass over the equator to minimise interference.

3.3.2 Governance Benefits of Satellite-Based Internet

The primary benefit of Internet connectivity is communication. Remote populations can communicate far more effectively with the rest of a country, or their leadership, and can alert others to a problem more quickly and accurately. For example, while a phone call may be able to tell someone that a sinkhole has occurred, a photo or video would show the absolute extent of such an issue, allowing a government to better react. Email enables detailed reports to be compiled and sent, while video calling supports detailed discussions. Developed countries already use these technologies thoroughly in practice, while most African countries simply cannot as there is no Internet backbone to support them.

The second benefit of the Internet to governance is education and awareness. Internet availability to the public not only increases standards of living but also education and awareness on current issues. Internet is taken for granted in first world countries, and heavily supports secondary and tertiary education. Video tutorials and online encyclopaedias have hundreds of thousands of viewers. There are full online university courses that people can access for low cost or free. And e-learning has been essential in developed countries during school lockdowns as a result of the current Covid 19 pandemic.

Africa is currently the least-educated region of the world by far. Figure 3.11 shows the 2010 comparative Education Index compiled by the UN. It uses adult literacy and years of education as metrics to determine education levels. Africa is the worst continent by far. Giving Internet access to remote African schools, libraries and civic centres will be sure to bolster the education of everyone it is available to. The simple act of using the Internet improves literacy. The exact impact that access to the Internet has on education is hard to measure but is surely massive.

The effect that education has on governance is well-known. There is a clear relationship between good education and good governance.[20] It is the difficulty of breaking into this cycle that hinders many African countries.

Figure 3.12 shows the results of a study that analysed the education level and

---

[18]Joanna Bailey, "SpaceX wants to launch more Starlink satellites… 30,000 of them", *Get Connected*, 2019 https://www.getconnected.aero/2019/10/spacex-starlink-30000-more/ (accessed 14 December 2019).

[19]Peter B. de Selding, "OneWeb Fails (At Least for Now) To Soothe Satellite Interference Fears", *Space News,* 2015, https://spacenews.com/oneweb-fails-at-least-for-now-to-soothe-satellite-interf erence-fears/ (accessed 14 December 2019).

[20]Maureen Lewis, "Governance in Education: Raising Performance", *World Bank*, 22 December 2009 (accessed 14 December 2019).

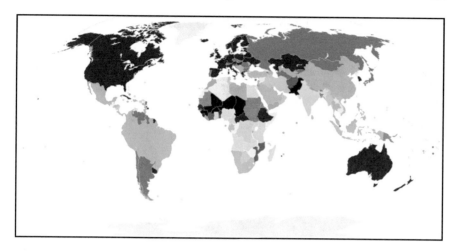

Fig. 3.11 Education Index of the world. Dark green is best, dark brown is worst (Tony0106 at English Wikipedia [CC BY 3.0 (https://creativecommons.org/licenses/by/3.0)], "Education Index Map", 2010, https://commons.wikimedia.org/wiki/File:Education_index_UN_HDR_200 8.svg (accessed 16 December 2019))

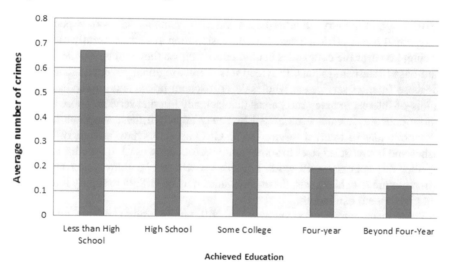

Fig. 3.12 Crime versus education level of US individuals (Raymond R. Swisher, Christopher R. Dennison, "Educational Pathways and Change in Crime Between Adolescence and Early Adulthood", Journal of Research in Crime and Delinquency, 2016, https://doi.org/10.1177/002242781 6645380)

crime reports of almost 15,000 individuals in the U.S.. Crime clearly decreases with education level significantly.

Education also improves the ability of the general population to report potential issues before they worsen and even solve problems on their own. Education indirectly makes some avenues of governance easier to achieve. Again, good governance is required to achieve good education as well.

As already said, access to the Internet can also increase awareness of current events. For example, it can alert people beforehand of possible disasters such as flooding, allowing them ample time to evacuate. It can also allow people to see the actions of their leaders, helping expose possible corruption. Of course, the Internet can be used to control a population through censorship and misinformation, but this is a necessary risk as the potential benefits of providing Internet to Africa are so great.

3.3.3 Potential Avenues to Implement Satellite-Based Internet in Africa

Existing geostationary satellite-based Internet remains far too expensive for widespread use in Africa. These future constellation systems currently have no set pricing, however the companies behind them promise they will be competitive with land-based Internet systems. Provided this is true, regional governments could bid for these Internet services at wholesale or discount prices and implement them at points-of-interest in many rural areas that lack any Internet service. Those countries whose economy cannot even afford that may approach these suppliers for charitable donations of Internet service (some GEO operators have already offered free dishes and low-cost services to some remote African towns[21]). It remains to be seen whether this approach will work, but since it will not cost any major land-based infrastructure to achieve this, these companies may be willing to supply limited (but still high-speed) bandwidth.

These mega-constellation systems will still require ground stations, albeit much smaller and easier to deploy than VSAT dishes, and will therefore need to be placed upon some structure where there is a clear view of the sky. This link could then connect to a traditional Wi-Fi router to allow access for normal devices. Governments could implement these connection points first at major libraries and schools in large towns that lack traditional Internet connections, and then provide for smaller and smaller settlements. Since so little infrastructure is needed, this could be expanded rapidly provided the funding is there. Government-owned clinics, police stations and offices could also provide Internet service to the surrounding area. Even if the main link provides an unlimited amount of data, they could, at first, limit the daily amount to each person to reduce abuse of the system.

---

[21] Kgaogelo Letsebe, "Konnect Africa offers rural community-based WiFi", *IT Web*, 2019, https://www.itweb.co.za/content/kxA9PO7Nk8EMo4J8 (accessed 16 December 2019).

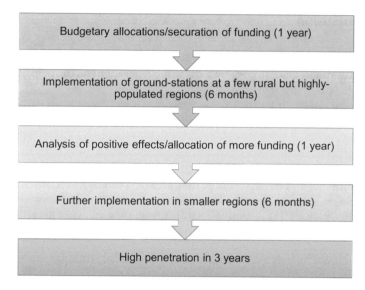

Fig. 3.13 Timescale of implementation of satellite-based internet

Governmental free Wi-Fi has already been introduced in some areas of South Africa with great positive effect. Ekurhuleni Metro of eastern Johannesburg has had a pilot programme that has notably allowed at least one person to start a successful business and aided many others in their studies.[22] South Africa, however, already has vast fibre systems near main centres, which most other African countries lack completely.[23]

Figure 3.13 describes a potential timeline for government provided free Wi-Fi. This is based on implementation timeframes of similar African infrastructure engineering projects. It may seem optimistic as most African governmental actions currently take far longer. However, the costs associated with this constellation-based Internet should be so low that it should be very easy to fit in a governmental budget. The install time is also very low, as it will just require one small antenna mounted on a rooftop and connected to a power source, rather than the kilometres of cable needed for a fibre connection. Hundreds of people could be given Internet access within the span of a few hours.

This, however, assumes that constellation Internet systems such as Starlink and OneWeb are successful. The technology does seem highly promising but remains unproven until the full network is established and in operation.

---

[22] "Wi-Fi changes life in Ekurhuleni", *South African Governmental News Agency* 2016, https://www.sanews.gov.za/south-africa/free-wi-fi-changes-life-ekurhuleni (accessed 16 December 2019).

[23] James Francis, "The Future of Fibre", *IT Web*, 4 March 2019.

3.4 Conclusion

The Internet and communication form a vital component of modern development and commerce. Its importance in governance is without question and can help governance indirectly in Africa in a myriad of ways.

Traditional satellite Internet remains costly with high latency but is still somewhat useful for African development. However, in the next decade constellation-based Internet such as Starlink and OneWeb will become operational, which will have the potential to greatly alter developmental activities in Africa.

Communication between communities in Africa will become better and enable digital governance and management of rural areas to become much better, as well as communication between leaders. The effect of governmental decisions may also be felt much faster, as well as emergency management. Education has the possibility of being greatly enhanced, as suddenly many rural children are exposed to the Internet and e-learning.

These possibilities are heavily dependent on how African governmental bodies implement such systems. With good planning, and provided the technologies are proven to work as intended, governance in Africa may be revolutionised in a short space of time with relatively low costs.

Chapter 4
Space Supporting Governance Through Applications for Human Development in Africa

Abstract This chapter focuses on the space applications that can be used to better human development and therefore governance in Africa. Sections such as defence, health, water management and biodiversity protection are examined with regards to the ability of space applications to support governance in these areas. Thereafter, some common toolsets are examined that will enable government officials to extract useful data for each of these areas. Special mention is made of ESA's programmes for development in Africa. Finally, the Sustainable Development Goals as mentioned in Chaps. 1 and 2 are modelled and ranked in terms of the benefit that space applications can provide for them.

4.1 Introduction

Monitoring of water resources is vital over Africa, to enable best use of this precious commodity. Until now reliable information has been difficult to access because of the high cost in equipment, manpower and communications, and because it is difficult to obtain these precious hydrological data from many countries.[1]

Prof. Berry (DMU's Earth and Planetary Remote Sensing Laboratory)

In its most basic terms, governance is the management of society by a group of people for the benefit of society. Extrapolating from this means that good governance requires both the protection of human rights and the sustainable and fair use of natural resources.

This chapter looks at how governance in specific areas can be aided by space applications. As discussed in Chap. 1, Africa's governance in many areas is still lacking. Traditional development of governance as occurred in more developed countries may take much longer in Africa given the vast and challenging landscape of Africa and its highly rural communities.

---

[1]ESA, Envisat tracking Africa's rivers and lakes to help manage water resources, 5 October 2005, https://www.esa.int/Applications/Observing_the_Earth/Envisat_tracking_Africa_s_rivers_and_lakes_to_help_manage_water_resources (accessed 20 Mai 2020).

A. Froehlich et al., *Space Supporting Africa*, Studies in Space Policy 28, https://doi.org/10.1007/978-3-030-52260-5_4

Fortunately, space applications can hasten this process, by providing a literal overview over occurrences in a region. Using different types of satellite data, from visible light to radar-based imagery as well as sophisticated programmes and algorithms that have improved significantly in recent years, governance can be performed more easily. This can enable Africa's governance to develop much faster than otherwise, especially when some of these resources are freely available.

This chapter will look at the space applications benefits for governance in the following areas:

- Peace, Security and Defence
- Health and Disease Control
- Water Management
- Biodiversity Management.

Thereafter, some common toolsets that may be used by governmental officials will be presented and described with possible uses and benefits. ESA has developed a number of projects with regards to development in Africa that will also be discussed, promoting better governance.

Finally based on the information contained in this publication, the SDGs (as mentioned in Chaps. 1 and 2) will be modelled and space's ability to support these goals will be examined.[2]

4.2   Peace, Security and Defence

As discussed in Chap. 1, Africa continues to face severe conflicts comprising border and civil wars as well as large-scale acts of violence and crime. EO (or remote sensing) satellites are a unique asset in promoting security.[3] Remote sensing can create many different types of images of both high spatial and temporal resolution. Furthermore, different sensing technologies in different bandwidths can detect many different things, allowing for a wide range of data and analyses.

This large amount of data is one of the major issues with remote sensing. People are simply unaware of the amount there is, or are insufficiently trained to use such a technology. Tchindjang et al. discuss this very issue, with the degradation of existing

---

[2]For additional information, see: I. Duvaux-Béchon, "The European Space Agency (ESA) and the United Nations 2030 SDG Goals", in *Embedding Space in African Society: The United Nations Sustainable Development Goals 2030 Supported by Space Applications*, ed. A. Froehlich (Cham: Springer, 2019), 223–236.

[3]See, for example: Gerald Hainzl, "Security in Africa: A Perception of Ongoing Developments", in *Embedding Space in African Society: The United Nations Sustainable Development Goals 2030 Supported by Space Applications*, ed. A. Froehlich (Cham: Springer, 2019), 255–264; and A. Froehlich, "Space Applications Supporting Justice", in *Embedding Space in African Society: The United Nations Sustainable Development Goals 2030 Supported by Space Applications*, ed. A. Froehlich (Cham: Springer, 2019), 265–279.

Fig. 4.1 Sentinel-2A image of Botswana (© Copernicus data 2016)

geographical information systems (GIS) training centres in Africa.[4] There is no hard evidence of the number of African militaries using remote sensing, but it is likely that it is very few. If this technology could be placed into the right hands, insurgencies could be detected and stopped in their infancy, kidnappings could be avoided, people recovered, and the general loss-of-life due to conflict in Africa could be reduced.

There are several different remote sensing technologies that can be used for security, the primary two of which will be examined in the following sections.

4.2.1   Visible Light Imagery

RGB or panchromatic satellite imagery produces images of light that humans are sensitive to, i.e. colours that we can see. Obviously, this imagery is better when taken in the daytime. This imagery can be used to identify insurgent camps and vehicles, provided they are not obstructed by trees or clouds.

There are a number of free sources of RGB satellite imagery. ESA's Sentinel-2 satellites provide free data of up to 10 m per pixel ground resolution (Fig. 4.1). NASA's Landsat 7 and 8 provide up to 15 m resolution of imagery, also for free

---

[4]Mesmin Tchindjang, Vincent Francis Menga, Jean Sylvestre Makak, Jean Pierre Nghonda, Marcellin Nziengui, Mesmin Edou, "Remote Sensing In Africa: Mapping Approach And Perspectives", *22nd International Cartographic Conference,* 2005.

Fig. 4.2 Landsat 8 image of Fort Collins, Colorado (NASA 2013)

(Fig. 4.2). While these are useful for applications that will be discussed further on, they are not high enough resolution for security applications.

There are however more and more providers that are offering high-resolution imagery for relatively low cost. One such company is Planet, that offers global daily imagery of 3.7 m resolution via its large fleet of Dove satellites or precision images up to twice a day of 0.72 m imagery using its SkySat spacecraft. Because of the nature of Planet's business structure, this imagery is the same price or less than many competitors, but is decided on a per-client basis. It is also worth noting that Planet has a large software frontend for its services, that allow it to be easily built into web-browsers and has automatic algorithms for detecting changes or deformations of buildings.

This makes Planet well suited to detecting mobilisation of insurgent vehicles or weapons, or the destruction or construction of strongholds and the continuous monitoring of them. While it will still cost governments a significant sum, it will be far cheaper than launching their own spy satellites (this will be discussed below).

Figure 4.3 is an image of the North Korean Chemicals Institute from SkySat. Notice the extremely high resolution, which can detect small details such as cars etc.

4.2.2 Synthetic Aperture Radar Imagery

Synthetic aperture radar (SAR) imagery is an active remote sensing technology. This means that the sun is not used to illuminate the target area but rather a microwave signal is sent from the satellite to reflect off the target and be collected by the satellite. The satellite then uses the time taken for the signal to return to create an image. SAR

Fig. 4.3 SkySat 0.72 m resolution image of North Korean Chemical Institute (Planet Labs, Inc [CC BY-SA 4.0 (https://creativecommons.org/licenses/by-sa/4.0)], 2018)

has a few advantages over visible light imagery. It works at night, through clouds, and even through some vegetation, depending on the wavelength of the radar used. These advantages are very useful to military applications where threats may be hidden by any of the above-described factors. Furthermore, any metal object on the Earth's surface acts as a good reflector for SAR, meaning that metal objects such as boats or military equipment are easily identifiable.

SAR is generally focussed around specific bandwidths with different applications. The wavelength usually corresponds to the level of detail required. The main bands of SAR are described in Table 4.1, with their major uses.

Table 4.1 Comparison of SAR bands

Band	Wavelength (cm)	Uses
P	30–100	Biomass estimations, experiments
L	15–30	High vegetation penetration, useful for soil or glacier observations
S	7.5–15	Rainfall and meteorological applications
C	3.75–7.5	Large object/maritime surveillance
X	2.4–3.75	Surveillance, building deformations

Fig. 4.4 Sentinel-1 SAR image of boats between Gibraltar and Algesiras (© Copernicus data 2017)

It should be noted that higher wavelength SAR is better at penetrating vegetation as the wavelength is too large to interact with leaves.

Sentinel-1 provides free C-band SAR imagery with normal 5 × 20 m resolution and up to 5 × 5 m in special cases. These images repeat about every 12 days. This data is suitable for larger military structures such as antennas etc. but is also useful to detect ocean vessels. It is already being employed by some African users such as the South African National Space Agency to detect illegal fishing vessels as well as identify possible pirates. Figure 4.4 shows a Sentinel-1 image with ships well-lit by SAR.

There are several SAR image providers from whom high-resolution data can be bought. TerraSAR-X is one such satellite that produces X-band images up to 1 m in resolution (Fig. 4.5), which can be used to identify individual land-vehicles. TerraSAR-X has a maximum revisit time of 11 days. These images do come at a price however, as the satellite needs to be focussed on specific areas.

4.2.2.1   SAR Imagery Disadvantages

SAR imagery requires far high levels of processing than RGB images, and as such require more training to use. They are inherently noisy and require filters to produce

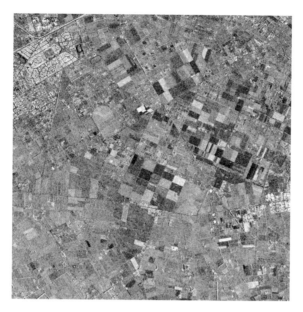

Fig. 4.5 False-colour TerraSAR-X image of Canterbury Plains, New Zealand. (c) DLR (Land care research [CC BY-SA 4.0 (https://creativecommons.org/licenses/by-sa/4.0)])

a clear image. If incorrectly handled, SAR images can quickly become incoherent. Furthermore, the inherent reflection of metals, which is in some ways an advantage, ends up obscuring other details. There are distortions due to the fact that SAR uses a side-angle view, making elevated areas appear to slope to one side.

4.2.3   Feasibility of Dedicated Defence Satellites in Africa

In 2014, Russia launched an S-band SAR reconnaissance satellite for South Africa (Kondor-E), at a cost of up to US$ 250 million to the South African defence department. This money was channelled through backdoors and was not part of the general budget allocations. This satellite is reportedly S-band at a maximum resolution of 1 m.[5] It was only designed for a 5 year operational lifetime and is therefore due to fail in the near future.[6]

Apart from some involvement in the Congo conflict, South Africa has had very limited involvement in military operations and does not have any major enemies.

---

[5] Anatoly Zak, "Russia orbits South-Africa's first spy satellite Kondor-E.", *Russian Space Web*, 2014, https://www.russianspaceweb.com/kondor-e.html (accessed 16 December 2019).

[6] David Todd, "South Africa purchases Kondor-E spysat from Russia just before its launch", *Seradata*, 4 December 2014.

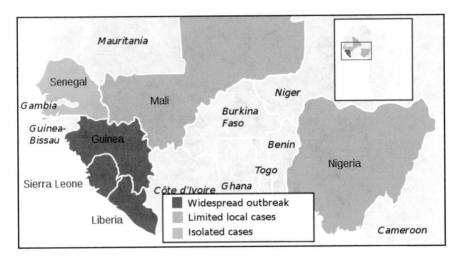

Fig. 4.6 Ebola epidemic of 2014 spread ("2014 Ebola Outbreak—detail timeline and projection", healthmap.org (accessed 16 December 2019))

Furthermore, TerraSAR-X provides as useful images for a far-smaller cost. Sentinel-1A was also launched in 2014 and provides for free adequate data for most possible South African uses of a spy satellite. With the limited lifetime, this satellite comes across as complete misuse of taxpayer funds and is possibly a form of major corruption.

With the small budgets that most African countries have, and the widely available third-party data, there is very little reason at present for any African country to build or buy dedicated spy satellites. Most warfare in Africa is unsophisticated, using old weapons from the more-developed regions of the world in guerrilla style insurgencies. Dedicated satellites built to find submarines or missile silos will be of little use in Africa, and any military that simply uses the data available on the Internet will be at a major advantage to the insurgencies that most likely do not.

4.3 Health and Disease Control

Due to Africa's vast and challenging terrain, as well as poverty and low living standards, disease epidemics are highly common in Africa. The West African Ebola epidemic of 2014 was disastrous, with about 30,000 deaths attributed to this virus (Fig. 4.6) shows the large land coverage of this outbreak.[7] Many people do not know

---

[7]"West African economies feeling ripple effects of Ebola, says UN", *United Nations Development Programme,* 12 March 2015.

that a second Ebola outbreak occurred in the DRC in 2018 and continues to the present day, with an estimated 4000 deaths so far.[8]

There are other highly dangerous diseases present in Africa, such as malaria and yellow fever. HIV/AIDS as well as tuberculosis are still highly common. Almost 8 million people lived with AIDS in South Africa alone as of 2018.[9]

While space technology cannot help with all diseases, it can provide significant support in terms of specific outbreaks in two major areas—communication and organisation, and vector tracking.

4.3.1 Communication, Organisation and Remote Treatment

As discussed above in the connectivity section, satellites can enable Internet connections worldwide, and as such may enable a remote village experiencing a certain disease outbreak to alert others. Doctors and health professionals can co-ordinate with those in nearby villages using satellite phones or the Internet, and map possible next areas of infection or routes that will allow relief.

Furthermore, with good enough bandwidth, telemedicine can be utilised, allowing experienced doctors from as far as the other side of the world to diagnose patients, prescribe medicine and even operate through a robot. This technology, known astelesurgery, is already present in some countries such as Kenya,[10] but is not as widespread as it could be. With machines costing around US$ 1 million each, it is little surprise. The technology is improving at a drastic pace, which gives hopes that African governments could afford to introduce these robots. For example, a device as simple as an iPad can be used to control the surgery.[11] Figure 4.7 shows the Da Vinci surgical system, which cost about US$ 2 million.

These systems are now adding 3D viewing, haptic feedback and fine motor adjustment to improve the ability of the surgeon to act as if he or she was in the room. General robotic surgery already minimises invasive activity and allows higher accuracy, which means that patients that receive telesurgery for certain procedures are at no disadvantage for the surgery. As soon as Internet bandwidth improves, possibly as a result of large constellation satellite systems, and the costs of these machines decrease, they can be more widely introduced in Africa, with the possibility of saving a great number of lives.

---

[8]"Ebola virus disease – Democratic Republic of the Congo. Disease Outbreak News, 14 September 2018", *World Health Organization (WHO)*, (accessed 16 December 2019).

[9]UNAIDS, unaids.org (accessed 16 December 2019).

[10]Chege Muigai, "Kenya: Health Services Go Online At Aga Khan With Launch of Telesurgery.", *Business Daily Nairobi*, 2011, https://allafrica.com/stories/201109081503.html (accessed 16 December 2019).

[11]Susie East, 2016. CNN. Doctor uses iPad to conduct remote surgery in Gaza. https://edition.cnn.com/2016/05/24/health/telesurgery-proximie-beirut-gaza/index.html.

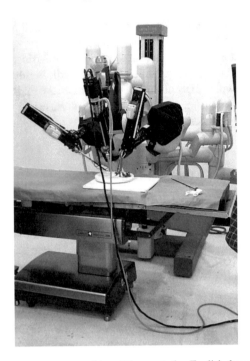

Fig. 4.7 da Vinci remote surgery machine (Nimur at the English language Wikipedia [CC BY-SA 3.0 (https://creativecommons.org/licenses/by-sa/3.0/)], "da Vinci Laproscopic Machine", 2006, https://commons.wikimedia.org/wiki/File:Laproscopic_Surgery_Robot.jpg (accessed 17 December 2019)

Table 4.2 Common disease vectors in Africa

Vector	Diseases
Mosquitoes	Dengue, malaria, West-Nile fever, Rift Valley fever
Ticks	African Tick Bite Fever
Sand-flies	Leishmaniasis, sand-fly fever
Midges	African horse sickness, bluetongue
Rats	Salmonella, plague

4.3.2 Vector Tracking and Prediction

Disease vectors are agents that transmit diseases from person-to-person.[12] Common vectors and the diseases they transmit in Africa are detailed in Table 4.2.

Obviously, these vectors are far too small for satellite cameras to be able to see. However, the conditions that are favourable to these vectors are easy to identify from

---

[12]Last, James, ed., (2001). A Dictionary of Epidemiology. New York: Oxford University Press. p. 185, ISBN 978-0-19-514,169-6. OCLC 207,797,812.

satellite data, and the risk of disease transmission when these conditions allow for high vector reproduction can be gauged with good accuracy from satellite data.

Some primary determining factors are soil moisture, atmospheric moisture and temperature, with mosquitos favouring warm and wet conditions for spread. ESA's Soil Moisture and Ocean Salinity (SMOS) satellite is one such spacecraft useful for determining vector spread. It orbits at 765 km and detects passive microwave radiation from the Earth that then, using interferometry (the difference in signals from different positions), determines the amount of moisture in near real-time. This lightly-processed data is freely available. MODIS is an infrared sensor on NASA's Terra satellite and provides freely available data of world surface temperatures in near real-time.

Using combinations of these types of data and some processing, the risk factors for mosquitos or other insects can be determined with high accuracy. ESA's VECMAP product is one such software tool that uses already processed data to indicate high-risk vector areas. This software is shown in Fig. 4.8.

While this software does have a licensing cost that varies according to the size of the organisation using it, the developers may be contacted for a discount for possible humanitarian reasons, especially in the most disease-stricken regions of Africa. Alternatively, an educational institution in Africa could develop its own software that can predict vectors in a similar manner and provide it free-of-charge to the areas that require it, since the required satellite data is freely available and only simple processing is required.

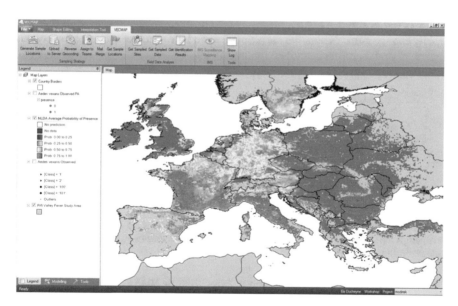

Fig. 4.8 VECMAP software showing vector risk in Europe ("VECMAP Software", https://business.esa.int/projects/vecmap (accessed 17 December 2019))

4.3.3   Microgravity Affecting Biology

Biology occurs differently in space, due to the lack of gravity that affects all biological processes on Earth. Consequently, experiments can be done in space to observe possible new treatment regimens for a wide variety of diseases The International Space Station (ISS) serves as a primary platform for these experiments and has already been involved in a high number of research studies beneficial to human health.

One such promising avenue is peptide therapy. Peptides are short-molecule proteins that occur naturally within the human body. Synthesised peptides have already been shown to be useful in the treatment of auto-immune diseases, infections and cancers, but are typically held back by short physiological lifetime, i.e. they do not last long in the body.[13] The Japanese Kibo module of the ISS is currently performing tests regarding synthetisation of these peptides in space. Because of the lack of gravity these peptides grow into non-standard shapes and last much longer in the human body. Once these drugs have been synthesised in space, they can be copied into techniques on the ground leading to large scale development of such treatments.[14]

This is just one of many space experiments to benefit human health, with many other ongoing experiments to develop vaccines, such as for salmonella, which is highly common in Africa,[15] and the development of biomedical technologies such as a robotic arm made for space that is now used to remove tumours.[16]

While these types of experiments cannot be directly used by governments to ensure better health in a region, they may lead to large-scale breakthroughs that will rapidly change Africa's health outlook, such as an HIV vaccine or a new tuberculosis treatment, and therefore cannot be ignored.

---

[13]Michael Johnson, "Finding the Keys in Space to Treat Diseases on Earth", *NASA*, 6 March 2019, https://www.nasa.gov/mission_pages/station/research/news/b4h-3rd/eds-space-research-to-treat-earth-diseases. (accessed 18 December 2019).

[14]Ibid.

[15]Tara Ruttley, "International Space Station Plays Role in Vaccine Development", *NASA*, 29 February 2012, https://www.nasa.gov/mission_pages/station/research/benefits/vaccine_develop ment.html (accessed 18 December 2019).

[16]Kristine Rainey, "neuroArm: Robotic Arms Lend a Healing Touch", *NASA*, 29 February 2012, https://www.nasa.gov/mission_pages/station/research/benefits/neuro_Arm.html (accessed 18 December 2019).

4.4 Water Management

One of the major governmental challenges for Africa is water management.[17] Africa faces far higher water scarcity than any other region of the world. It is estimated that by 2025, half of Africa's countries will face severe water stress.[18]

There are several reasons for this. The first is simply the geographical nature of the continent. Africa does not contain many large river or lake systems. The Sahara itself is 30% of Africa's total surface area.[19]

Second, this natural tendency to be a dry region is exacerbated by the effects of climate change (which will be discussed in itself in a later section), with major droughts becoming increasingly common throughout the 2010s. In 2011, East Africa faced a drought so severe that it is believed that up to 260,000 people died because of the resulting famine and disease.[20] Another drought in 2019 among eastern and southern African countries has left an estimated 45 million people facing food shortages.[21]

The effects of climate change are predicted to become worse, and Africa's population is growing extremely quickly, having almost doubled in the past 20 years,[22] which means that water systems are becoming even more strained. This, coupled with the fact that Africa lacks adequate infrastructure to both store and purify water, means that the continent will be facing even more stress in the near future with a possible massive death toll unless measures are taken to mitigate these droughts.

Space technology offers a large opportunity to help in preventing these water stresses. The remainder of this section will examine several different ways in which satellites can and have helped mitigate water scarcity. These ideas will be based on a report compiled by University of Cape Town's SpaceLab in direct response to the Cape Water Crisis of 2017–2019 in South Africa.

---

[17] For additional information related to Tunisia, see: W. Abdallah, M. Allani, R. Mezzi, et al., "A Contribution to an Advisory Plan for Integrated Irrigation Water Management at Sidi Saad Dam System (Central Tunisia): From Research to Operational Support", in *Embedding Space in African Society: The United Nations Sustainable Development Goals 2030 Supported by Space Applications*, ed. A. Froehlich (Cham: Springer, 2019), 65–80.

[18] Christopher W. Tatlock, "Water Stress in Sub-Saharan Africa", *Council on Foreign Relations*, 3 August 2006, https://www.cfr.org/backgrounder/water-stress-sub-saharan-africa (accessed 18 December 2019).

[19] Cook, Kerry H., Vizy, Edward K, "Detection and Analysis of an Amplified Warming of the Sahara Desert", Journal of Climate 28 (2016), 6560.

[20] Associated Press, "Famine Toll in 2011 Was Larger Than Previously Reported", *The New York Times,* 29 April 2013.

[21] Obi Anyadike, "Drought in Africa leaves 45 million in need across 14 countries", *The New Humanitarian,* 10 June 2019, https://www.thenewhumanitarian.org/analysis/2019/06/10/drought-africa-2019-45-million-in-need (accessed 18 December 2019).

[22] "Population of Africa (2019) – Worldometers". www.worldometers.info (accessed 18 December 2019).

Fig. 4.9 Sentinel-2B image of Theewaterskloof dam in Cape Town at 12% capacity in February 2018. © Copernicus (2018) (Antti Lipponen [CC BY 2.0 (https://creativecommons.org/licenses/by/2.0)], "Theewaterskloof Dam", 11 February 2018, https://commons.wikimedia.org/wiki/File:Theewaterskloof_Dam_2018_02_10_(28425520089).jpg (accessed 18 December 2019))

4.4.1   Remote Sensing for Water Management

Bodies of water can easily be identified and monitored from space not only for the amount of water but also its quality. Hydrology network maps are a commonplace product of Earth Observation, which can accurately show changes in a water system from the original Landsat images[23] up to the present time. These optical images can quickly and effectively show how larger open bodies of water have been affected by mismanagement or climate change. Smaller bodies of water such as farm reservoirs require higher resolution images, but these are far more commonplace now than they have been before due to the likes of Planet and other high-resolution satellite operators. Figure 4.9 clearly shows the effects of a drought on Cape Town's main supply dam in 2018.

SAR can also be used to augment these hydrological maps. SAR has two major advantages over visible light data, which can only show unobscured bodies of water in size. The first is penetration. Water is clearly visible in SAR data due to the way that the water surface interacts with the radar. Since SAR is penetrating, it can be

---

[23]"Landsat Program Chronology", *NASA*, 2 December 2016.

Fig. 4.10 SAR image of flooding of Po river in Italy. © Copernicus (2019) (ESA [CC BY SA 3.0 https://creativecommons.org/licenses/by-sa/3.0/za/], 26 November 2019, https://earth.esa.int/web/guest/missions/esa-operational-eo-missions/sentinel-1/news/-/article/floods-in-northern-italy (accessed 18 December 2019))

used to monitor rivers, lakes and dams obscured by vegetation cover or clouds during extended rainy seasons. Figure 4.10 shows a false colour Sentinel-1 image of flooding in Italy, clearly showing how water can be highlighted by SAR.

Furthermore, land elevation can be measured through either interferometry (the time difference between two SAR satellites measuring the same area, Sentinel-1 is capable of this[24]) or through a single radar signal sent directly below the spacecraft to measure altitude such as the now-ended Topex mission.[25] Interferometry has largely replaced radar altimetry as it is more accurate. These techniques can be used to produce digital elevation maps (DEMs) which can show the current altitude of water bodies or interferograms which show rapid changes in water and/or land height. Figure 4.11 shows one of these DEMs of the entire world created by NASA. This

---

[24]ESA, "Interferometry", https://sentinel.esa.int/web/sentinel/user-guides/sentinel-1-sar/product-overview/interferometry (accessed 18 December 2019).

[25]"Ocean Surface Topography from Space", *NASA/JPL*, Archived from the original on 2001–10-23 (accessed 18 December 2019).

Fig. 4.11 NASA heightmap of the world including water levels

image can serve as a reference point to determine all future ocean and lake height levels.

These techniques clearly allow surface water level monitoring with very high temporal resolution. Satellite data can also show the quality of water, not only the quantity. One such useful data point is the amount of chlorophyll-a near the water surface. This directly corresponds to the amount of phytoplankton in the water, and in turn how nutrient rich it may be for fish and, hence, overall water quality. This requires simple processing in the blue bands of visible light which, for example, NASA's MODIS can do.[26] Figure 4.12 shows the amount of chlorophyll-a in both the ocean and large water bodies in central and southern Africa using MODIS data.

Satellites are also good at imaging algal blooms.[27] While many algal blooms may be harmless, some may be comprised of harmful plankton or cyanobacteria that create toxins harmful to water or land-based life.[28] These are known as harmful algal blooms (HABs) and while most often occurring in the sea, they may occur in larger lakes and rivers and result in the poisoning of drinking water. Many different satellites have been able to detect algal blooms. The best performing satellite system at present to image them is the Sentinel-3 group, which can detect algal blooms as

---

[26] "MODIS Chlorophyll-a Concentration", *NASA*, https://modis.gsfc.nasa.gov/data/dataprod/chlor_a.php (accessed 22 December 2019).

[27] For a related discussion on imaging, see: M.A. Tajelsir Raoof, D. Fritsch, and R. Abdalla, "Signal Coverage of Low-Land Areas Using Geographic Information Systems, Case Study: Kassingar Area, Sudan", in *Space Fostering African Societies: Developing the African Continent through Space, Part 1*, ed. A. Froehlich (Cham: Springer, 2020), 219–230.

[28] Stewart I, Seawright AA, Shaw GR, "Cyanobacterial poisoning in livestock, wild mammals and birds – an overview." Cyanobacterial Harmful Algal Blooms: State of the Science and Research Needs. Advances in Experimental Medicine and Biology (2008). 619. pp. 613–637.

Fig. 4.12 Chlorophyll-a concentration in October 2019 ("NASA Worldview", *NASA*, 25 October 2019, https://go.nasa.gov/36uL0Rs (accessed 22 December 2019))

small as 300 m^2.[29] Figure 4.13 shows a Sentinel-3 image of an HAB in Lake Erie in North America. The bright green strands are the cyanobacteria.

There are some specific indices as well for measuring the exact level of algae, such as the surface algal bloom index (SABi) that performs a calculation based on the relative intensities of blue and green bands of the sensor.[30] This provides an accurate baseline to compare the magnitude of blooms to each other.

Outside of hydrology mapping and water level and quality measurement, remote sensing also enables the determination of settlement characteristics with regards to water sources. In other words, it can be used to predict which settlements are going to be vulnerable to flooding or droughts, and which might rapidly expand and create pressure on a water source. The Global Urban Footprint (GUF) programme is one such data source, which combines census information with TerraSAR-X and TanDEM-X data to accurately map all human settlements in the world. Due to the materials used in human construction as well as the geometric shapes, buildings can be easily separated from natural features in SAR imagery.[31] Figure 4.14 shows a GUF map of Delhi in India.

Of course, water primarily affects vegetation, both natural and crops. Remote sensing is extremely effective at judging vegetation density. It was found that leaves

---

[29]"Data application of the month: Harmful Algal Blooms", *UN-SPIDER*, https://www.un-spider.org/links-and-resources/data-sources/daotm-HABs (accessed 22 December 2019).

[30]"Data application of the month: Harmful Algal Blooms", *UN-SPIDER*, https://www.un-spider.org/links-and-resources/data-sources/daotm-HABs (accessed 22 December 2019).

[31]DLR, "Global Urban Footprint", https://www.dlr.de/eoc/en/desktopdefault.aspx/tabid-9628/16557_read-40454/ (accessed 22 December 2019).

Fig. 4.13 Cyanobacteria Algal Bloom of Lake Erie in North America in 2017. © Copernicus (2017) (Andy, "Algae Starting To Bloom", *Pixalytics*, 23 August 2017, https://www.pixalytics.com/algae-starting-to-bloom/ (accessed 22 December 2019))

Fig. 4.14 GUF of Settlements in Delhi, India ("New map offers precise snapshot of human life on Earth", *ESA*, 18 November 2016, https://www.esa.int/Applications/Observing_the_Earth/New_map_offers_precise_snapshot_of_h uman_life_on_Earth (accessed 22 December 2019)) © DLR

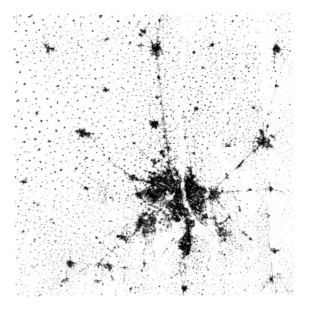

are very good at reflecting near infrared frequencies and, as such, processing techniques using visible light and near-infrared satellite imagery were developed to effectively measure vegetation density.

Perhaps the most common of these techniques is the Normalised Difference Vegetation Index (NDVI) that stemmed from a NASA study in the early 1970s.[32] NDVI is described by Eq. 4.1.

$$NDVI = \frac{NIR - Red}{NIR + Red} \tag{4.1}$$

"NIR" and "Red" are the relative intensities of the light received by the satellite corresponding to those spectra. It produces a result from -1 to 1. For dense, green vegetation, it produces a result of close to 1. Clouds and snow will correspond to a value of up to -1. Soil and urban areas will be around 0.[33] This processed information can then be placed on top of a map or satellite image. Landsat,[34] Sentinel-2[35] and MODIS[36] all have this capability. Figure 4.15 shows the NDVI for Africa in October 2019 based on MODIS data. Notice the lack of vegetation in the large desert areas.

Figure 4.15 was derived for free using NASA Worldview,[37] an online tool providing primarily MODIS and Landsat data as well as some other sources. It provides up-to-date data and enables animations within the browser to determine changes on the Earth. Because the NDVI is a simple process using free data such as that on NASA Worldview, it will be easy for governments to take advantage of and see what regions may be experiencing droughts early on and determine where crops might fail in order to bolster food supplies before a major issue occurs. Vegetation indices cannot solve the problem of drought, but they may give important early warnings.

4.4.2 Hydrological Modelling

It is well known that satellite images can be used to predict weather, as cloud systems can be clearly seen forming and approaching regions before they reach full strength. This can help lead to evacuations before cyclones or flooding. Figure 4.16 shows

---

[32] Rouse, J.W, Haas, R.H., Scheel, J.A., and Deering, D.W., "Monitoring Vegetation Systems in the Great Plains with ERTS", Proceedings, 3rd Earth Resource Technology Satellite (ERTS) Symposium (1974), vol. 1, pp. 48–62. https://ntrs.nasa.gov/archive/nasa/casi.ntrs.nasa.gov/19740022592. pdf (accessed 22 December 2019).

[33] Ibid.

[34] NASA, "Landsat Data Continuity Mission Brochure."

[35] "MultiSpectral Instrument (MSI) Overview", *Sentinel Online European Space Agency.*

[36] "NASA Worldview", *NASA,* 27 October 2019, https://go.nasa.gov/2PRPue2 (accessed 22 December 2019).

[37] Ibid.

Fig. 4.15 MODIS derived
NDVI for Africa, 29 October
2019 (Ibid.)

Fig. 4.16 Hurricane Katrina
("Hurricane Katrina at peak
intensity—MODIS", NASA,
28 August 2005, https://en.
wikipedia.org/wiki/Hurric
ane_Katrina#/media/File:
Hurricane_Katrina_August_
28_2005_NASA.jpg
(accessed 22 December
2019))

Hurricane Katrina imaged by MODIS at its peak intensity. Its path and windspeeds
could be predicted by satellite images like these.

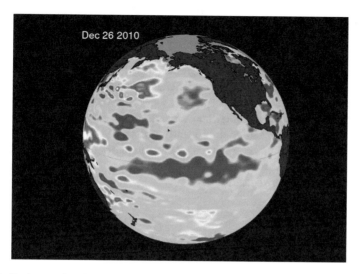

Fig. 4.17 Cool sea surface during a La Niña event in 2010 ("NOAA Jason-2 sea surface temperature monitoring—strong La Niña," *NASA/JPL*, 2011, https://www.nasa.gov/topics/earth/features/strong-la-nina.html (accessed 22 December 2019))

While short-term weather systems can be predicted fairly accurately, long-term rainfall modelling is much harder to perform. Rain gauges located at many meteorological stations around the world have been measuring rainfall for many decades and have managed to discover some patterns. However, big changes to the climate system such as El Niño and La Niña, which lead to periods of excess rainfall and drought respectively, are still not well understood. They occur due to the reverse of trade winds, but it is not known what causes these winds to reverse. They are known to occur every 2–7 years but that is a very wide range of time.[38] They strongly correlate to sea surface temperatures but again cannot be predicted. Figure 4.17 shows the cool sea-surface temperature during a strong La Niña event, using an infrared sensor on the Jason-2 satellite.

Global climate events such as El Niño and La Niña, as well as climate change, make it extremely hard to model long term rainfall and predict droughts far into the future. There is a promising avenue of research however, that combines machine learning with the large data sources that remote sensing provides.

---

[38]Mike Carlowicz and Stephanie Schollaert Uz, "El Niño", *NASA Earth Observatory,* 14 February 2017, https://earthobservatory.nasa.gov/features/ElNino (accessed 22 December 2019).

Machine learning (AI) allows abstract interpretation of data sources such as images with recognition in some categories beating human abilities.[39] What this means is that if an artificial neural network to detect climate change is sufficiently developed and trained on previous data, it may be able to recognise extremely subtle patterns that humans may be oblivious to. Scientists have already attempted this with some success.[40]

As both neural network design and datasets improve, the prediction ability of these AI models should improve significantly, possibly to the point where the size of a cyclone or drought could be predicted several months or even years into the future. This will be a true innovation in the use of satellite data to manage water and is likely to happen in the next few decades.

4.4.3   Implementation of Water Management for African Governments

While satellite data is very useful in terms of prediction and monitoring, it cannot physically alter conditions on the ground. Africa still heavily lacks basic water infrastructure. Dams, piping and water treatment plants simply do not exist throughout most of the continent.[41] These basic infrastructures need to be introduced first.

The success of these types of infrastructures has two main requirements. The first is capital. The second is skills and training to maintain water infrastructure after it has been built, which Africa still has a massive shortage of.[42] Wikipedia has a list of many charities and NGO's that provide water infrastructure in Africa.[43]

[39]Sissi Cao, "Google's DeepMind AI Beats Humans Again—This Time By Deciphering Ancient Greek Text", *Observer*, 21 October 2019, https://observer.com/2019/10/google-deepmind-ai-mac hine-learning-beat-human-ancient-greek-text-prediction/ (accessed 22 December 2019).

[40]Rhee, Jinyoung & Im, Jungho & Park, S., "Drought Forecasting Based On Machine Learning Of Remote Sensing And Long-Range Forecast Data. Isprs", International Archives of the Photogrammetry, Remote Sensing and Spatial Information Sciences (2016), XLI-B8. 157–158. 10.5194/isprsarchives-XLI-B8-157–2016.

[41]"Water Infrastructure for Economic Growth", *African Minister's Council on Water,* 2016, https://www.amcow-online.org/index.php?option=com_content&view=article&id=148&Ite mid=78&lang=en (accessed 22 December 2019).

[42]Fiona Harvey, "Africa's shortage of engineering skills 'will stunt its growth'", *The Guardian,* 14 September 2016, https://www.theguardian.com/global-development/2016/sep/14/africa-shortage-of-engineering-skills-and-female-students-will-stunt-its-growth (accessed 22 December 2019).

[43]"List of water-related charities", *Wikipedia,* https://en.wikipedia.org/wiki/List_of_water-rel ated_charities (accessed 22 December 2019).

• ACTS	• Office International de l'Eau
• 28Bold	• Ok Clean Water Project
• The Ann Campana Judge Foundation	• One Drop
• Blood: Water Mission	• Poured Out Inc
• Blue Planet Network	• Project Water for LifePump Aid-Water for Life
• CARE	• Pure Water for the World
• CAWST: Centre for Affordable Water and Sanitation Technology	• Relief Network Ministries, Inc. (RNM)
• Charity: water	• Ryan's Well Foundation
• Circle of Blue	• Thirst Relief International
• Clean Water for Haiti DigDeep (Africa)	• UNICEF—WASH
• DIGDEEP Water	• Voss Foundation
• Drop in the Bucket	• WASH Advocates
• The Gender and Water Alliance	• Water Access Now
• Generosity Water	• Water and Sanitation Program
• Generosity.org	• Water For People
• Global Water	• Water is Basic
• Global Water Foundation	• Water Is Life
• H_2O For Life	• The Water Project
• Healing Waters International	• Water to Thrive
• Hope Spring	• The Water Trust
• Initiative: Eau	• Water Wells For Africa
• International Medical Corps	• Water.org
• IRC	• WaterAid
• Just a Drop	• WaterCan/EauVive
• The Last Well	• Well Constructed
• Lifewater Canada	• Well Aware
• Lifewater International	• Wells Bring Hope
• Living Water International	• Wells of Life
• The Millennium Water Alliance	• World Vision

The issue with almost all of these projects is that they are for small-scale wells or filtration systems. For this they cannot be criticised as large-scale water infrastructure such as dams and pipelines are extremely costly and are usually only reserved to government budgets.

There is another actor however that has been building dams in Africa. China is building a number of dams in Africa including the world's largest dam.[44] These dams will mainly be hydropower for China's own resource operations, and while they will provide power for some regions, they are likely not being built for the benefit of most indigenous people, with one such dam already having displaced 50,000 people.[45]

The fundamental solution therefore appears to be one of medium-scale. Suitable locations need to be found for smaller dams and water treatment facilities that African governments can afford to build and maintain, and this is where the true practicality of satellite data may be useful to African water governance. Using different layers of satellite information, namely land use, elevation, groundwater level, rainfall amounts and settlement distribution, useful data will appear on the best locations for new smaller dams or other water infrastructure. Free software such as QGIS[46] or the SNAP toolbox[47] enable this type of processing relatively easily. However, dams face evaporation as well as shortages due to drought, and therefore cannot be thought of as a total solution to Africa's water problems. Fortunately, satellite technology can assist in treating other symptoms of drought and even some causes.

4.4.3.1   Rooftop Catchment

One such solution is rooftop catchment, which of course applies more to urban areas than rural settlements. Satellite imagery can accurately gauge the rooftop size of many buildings and, using rainfall models, can quickly identify the amount of water that could be obtained through these methods.

OpenStreetMap (OSM)[48] can be used to do this effectively, where rooftops can or have already been drawn and can be averaged to determine rooftop area. Combining this with historical rainfall amounts will allow complete possible water catchment amounts of a city to be identified. Figure 4.18 shows the default OSM map of Johannesburg CBD.

Most African regions do not treat their wastewater, which remains a huge problem.[49] These rooftop catchment systems could supplement this loss in a cost-effective way. A smaller building could collect up to 3000 L at a time for a cost of

---

[44]"China building 'world's biggest dam' in Africa – and SA will benefit", *Cape Business News,* 7 August 2019, https://www.cbn.co.za/featured/china-building-worlds-biggest-dam-in-africa-and-sa-will-benefit/ (accessed 26 December 2019).

[45]"Chinese Dams in Africa", *International Rivers,* https://www.internationalrivers.org/campaigns/chinese-dams-in-africa (accessed 26 December 2019).

[46]"QGIS", https://qgis.org/en/site/ (accessed 26 December 2019).

[47]"SNAP Toolbox", https://step.esa.int/main/toolboxes/snap/ (accessed 26 December 2019).

[48]"Open Street Maps", www.osm.org (accessed 26 December 2019).

[49]Christoph Haushofer, "Africa: The Reuse Of Treated Wastewater For Drinking Water", 21 March 2019, https://www.afrik21.africa/en/africa-the-reuse-of-treated-wastewater-for-drinking-water/ (accessed 4 January 2020).

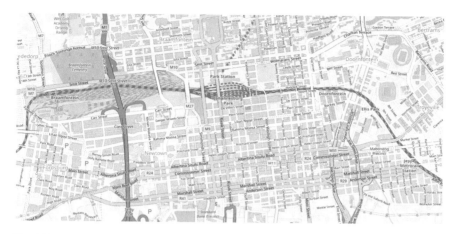

Fig. 4.18   OSM of Johannesburg CBD. © OpenStreetMap contributors (Ibid.)

less than US $2000.[50] With an average per-person use of 47 L a day in Africa,[51] this could supply a family of four with water for two weeks. Alternatively, in times of high water crisis, it could supply drinking water for many months. Using urban catchment area modelling can help to accurately size water tanks and predict refill times, allowing financial justification of the costs involved.

4.4.3.2   Vegetation Affecting Water Management

It is well known that drought affects vegetation, both natural and crops, often leading to immature deaths of many plants and an inability to grow new seedlings, leading to severe food shortages as well as increased chances of mass wildfires.

Vegetation, however, also affects water in two primary ways. Specifically in Africa, the first is invasive species consuming huge amounts of water. These invasive species came from areas with much higher amounts of rainfall such as Europe and North America and have managed to dominate growth in Africa. The average pine tree uses about 25 L of water a day,[52] more than many people in Africa. While these trees do transpire some of this water into the atmosphere as water vapour, it is usually in insufficient amounts to form clouds, which along with African winds and heat, move this water vapour elsewhere. This effectively means that these trees are simply wasting this water, preventing it from reaching critical rivers and dams in the catchment areas and significantly lowering the availability of water to users in a

---

[50]Matthew Firth, "How Much Does A Water Collection System Cost?", *Enduraplas,* 13 March 2018, https://blog.enduraplas.com/water-storage-rain-harvesting/how-much-does-a-water-collection-system-cost (accessed 4 January 2020).

[51]"Water Consumption in Africa", *Water for Africa Institute,* 2002, https://water-for-africa.org/en/water-consumption/articles/water-consumption-in-africa.html (accessed 4 January 2020).

[52]"Forestry", *Sabie,* https://www.sabie.co.za/about/forestry/ (accessed 4 January 2020).

Fig. 4.19 Gum Tree Forest at OR Tambo International Airport in South Africa. © CNES/Airbus ("Google Maps," https://www.google.com/maps/@-26.1515499,28.228137 8,3804m/data=!3m1!1e3 (accessed 4 January 2020))

region. There are many different types of invasive plant species affecting Africa, but there is good news in this. High-resolution satellite imagery allows easy identification of these species, both by manual verification and by machine learning. Furthermore, it can identify forests nearest to critical catchment areas and show which will be easiest to remove. This lumber can then further be used as a construction material etc., meaning that there will be almost total benefit in removal of such invasive forests.

Google Earth is useful in this as a free tool. Figure 4.19 shows a gum tree forest at OR Tambo Airport in Johannesburg South Africa. Gum trees are indigenous to Australia which has significantly different climate attributes to South Africa, and also use extremely large amounts of water. These trees are easily and quickly identifiable at this resolution due to their dark green colour and density, when compared with native South African species. A machine learning algorithm could be developed to automatically identify them. A researcher could easily identify a number of problematic invasive forests around rivers and dams in a few hours and then work with other government or private entities to plan removal of such forests, with the removal of a few thousand trees being able to save millions of litres of water in a year. This process was identified and preliminarily discussed in "Space Fostering African Societies" published by Springer.[53]

---

[53]Nicolas Ringas and James Wilson et. al., "Dry the beloved country", in: Annette Froehlich (ed.) Space Fostering African Societies - Developing the African Continent through Space, Part 1, Southern Space Studies, 2020, pp. 273–331.

Fig. 4.20 Flash Flood in KwaZulu-Natal, South Africa in April 2019 (Aanu Adeoye, "70 people killed in South Africa floods", *CNN*, https://edition.cnn.com/2019/04/24/africa/51-dead-south-afr ica-flood-intl/index.html (accessed 4 January 2020))

The second way that vegetation can affect water use is through run-off amounts. Areas that have experienced desertification lose natural scrub cover that protects extreme run-off of water after heavy rains. This leads to a first issue of flash-flooding endangering people, animals and property, and a second issue of this water leaving the region where it is needed and entering the sea rather than the ground. Figure 4.20 shows the destruction from a flash-flood that occurred near Durban in South Africa in 2019. 70 people lost their lives to this flooding, which highlights the importance of mitigating such tragic events.

Vegetation physically slows down the travel of water giving it time to seep into the ground. It also loosens soil allowing water to enter the ground more easily. The loss of this ground cover vegetation is extremely detrimental to the control of water after heavy rains.

Satellite imagery aids in the identification of these regions in much the same way as it does in detecting alien trees. Ground can be very easily identified as being covered by vegetation or barren, and this can be used effectively near key settlements in dry areas to either try to regrow natural vegetation, or to use artificial methods to prevent high run-off such as checking dams and drainage ditches. Satellite images can significantly decrease the necessary planning times, and with the number of floods set to increase due to climate change,[54] this is highly important.

---

[54]Robin McKie, "Global heating to inflict more droughts on Africa as well as floods", *The Guardian*, 16 June 2019, https://www.theguardian.com/science/2019/jun/14/africa-global-heating-more-dro ughts-and-flooding-threat (accessed 4 January 2020).

4.5 Satellite Technology for Wildlife and Biodiversity Protection

Wildlife directly contributes huge amounts of wealth to Africa. Many regions survive only on tourism for the vast biodiversity that Africa offers. It is lamentable that it is estimated that up-to half of all African species face extinction, either directly due to poaching or due to other human causes.[55] It is imperative that governments try to protect this biodiversity, as it contributes directly to the African economy through tourism—Africa received US$ 37 billion in 2017 from tourism alone, with further exponential growth expected in the near future.[56]

Furthermore, Africa's complex biodiversity maintains an essential balance in natural cycles. With so many communities relying purely on subsistence methods, any extinction, be it of a plant, insect or large animal such as a rhino, may have far reaching consequences for such communities.

This section will examine methods of protecting Africa's wildlife and biodiversity through satellite technology, which directly contributes economically and therefore indirectly to governance.

4.5.1 Wildlife Tracking

Wildlife tracking is primarily useful for two reasons. The first is to observe population counts and migration routes and the effects that loss of natural habitat and increased urbanisation may have on them. The second is to track certain endangered species to directly protect them from poachers.

Rhinoceroses are a prime target for poachers throughout Africa, due to the miscon-strued and completely incorrect belief in Asia that rhino horn is some form of wonder medicine. The Northern White Rhino is believed to have become extinct in 2018, following the death of the last known male.[57] Rhino horn is composed of exactly the same proteins as human nails and offers no health benefits.

The Southern White Rhino and Black Rhino still face huge threats, with more than one thousand poached each year in South Africa alone.[58] Many rhinos are de-horned (Fig. 4.21) which is painful for them and then are still killed for any piece of

---

[55]"Half of African species face extinction", *BBC News*, 23 March 2018, https://www.bbc.com/news/world-43516211 (accessed 4 January 2020).

[56]David Monyae, "Africa needs to make most of tourism surge", *IOL*, 14 August 2019, https://www.iol.co.za/news/opinion/africa-needs-to-make-most-of-tourism-surge-30763596 (accessed 4 January 2020).

[57]Sarah Gibbens, "After Last Male's Death, Is the Northern White Rhino Doomed?", *National Geographic*, 19 March 2018, https://www.nationalgeographic.com/news/2018/03/northern-white-rhino-male-sudan-death-extinction-spd/ (accessed 4 January 2020).

[58]"Poaching Stats", *Save The Rhino*, https://www.savetherhino.org/rhino-info/poaching-stats/ (accessed 6 January 2020).

Fig. 4.21 Dehorned rhino in the Kruger National Park, South Africa (Bernard DuPont (CC BY-SA 2.0), "Dehorned Rhino" (accessed 6 January 2020))

remaining horn. Armed patrols to protect rhinos have been implemented in recent years, with exchanges of gunfire between the poachers and protectors. While most of the poachers are African citizens, they usually work for larger organisations that may be governmental or even internationally-based.[59] The cost of protecting rhinos is astronomically high, and some human lives have been lost in the effort. While rhinos may be large animals, they are still difficult to track in dense bush and can be a struggle for those protecting them to find.

Fortunately, a satellite system known as Argos[60] allows rapid tracking of wildlife of all types. Argos is based on the doppler active tracking system that was developed before the creation of GNSS systems. Unlike GNSS, where only the receiver knows its location (i.e. passive), doppler systems involve two-way communication which allows the position of a receiver to be seen anywhere else in the world. The Argos system allows locations to be observed approximately once every two hours. They are accurate to 150 m but require a second-pass before true-location can be discovered due to the bilateral nature of the Doppler effect. However, modern trackers use GNSS also to enhance accuracy, giving location within 10 m without requiring a second pass.

Of course, Argos trackers have limited battery life depending on whether they use GPS and what size they are. However, many are designed to transmit more than

[59]Bryan Christy, "Special Investigation: Inside the Deadly Rhino Horn Trade", *National Geographic,* October 2016, https://www.nationalgeographic.com/magazine/2016/10/dark-world-of-the-rhino-horn-trade/ (accessed 6 January 2020).

[60]"ARGOS", argos-system.org (accessed 6 January 2020).

Fig. 4.22 Emperor Penguin fitted with Argos tracker ("King penguin fitted with an Argos transmitter", *Carbon Brief*, https://www.carbonbrief.org/climate-change-a-serious-threat-to-king-penguins-study-warns/penguin-four)

8 times a day and have a battery life of over 2 years.[61] Argos also scales very well. Rhinos are large and may carry a large transceiver, but some Argos trackers are small enough for even small birds to carry. Argos is also not-for-profit, and therefore costs are very low compared to the costs of having armed guards for specific wildlife. Figure 4.22 shows an Emperor penguin fitted with a lightweight Argos tracker.

There are of course other methods to track wildlife, but Argos remains the most cost-effective solution for where GSM networks or VHF/UHF radios cannot reach.

4.5.2   Habitat Monitoring

Using many of the same Earth observation methods as described in previous sections, habitat and biodiversity health can be easily monitored by many free and easily-available satellite image sources. A prime example of the use of this would be to monitor the destruction of rainforest in central Africa, a home to one of the

---

[61] Telonics, Quality Electronics for Wildlife, Environmental Research, and Special Applications https://www.telonics.com/products/gps4/gps-argos.php (accessed 6 January 2020).

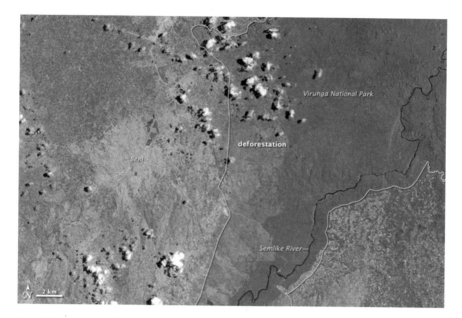

Fig. 4.23 Agricultural land bordering forest in the DRC ("Gorillas in the Midst of Extinction", *NASA*, January 2005 (accessed 6 January 2020))

rarest species on Earth, the gorilla. Most of this forest destruction is barely legal and performed by companies that mostly have no concern for local populations. Figure 4.23 shows deforestation caused by agriculture. The plight of the gorillas is only increased further by poaching for traditional medicines.

Other species in Africa also have highly-specialised habitat requirements. African wild dogs are the most intelligent African predator and have the most intricate social systems of any larger African mammal. There are only about 6,600 African wild dogs left,[62] mainly because a single pack may require up to 1000 km^2.[63] Urbanisation and fragmentation has significantly decreased their numbers.

Satellite data allows tracking and monitoring of deforestation to within a day, enabling illegal habitat destruction to be quickly identified and acted upon, as well as predicting the short and long-term effects of such habitat removal. Furthermore, it can play a role in eco-friendly development planning, allowing new developments to minimise impact on habitats, while still promoting economic growth within the region.

This data from common free sources is already being used by a few private organisations in Africa to identify habitat destruction. However, governments and larger bodies will need to act on this data to be able to prevent widespread habitat destruction such as that of gorillas in central African jungles.

---

[62]"Lycaon pictus", *IUCN Red List of Threatened Species.*

[63]"African Wild Dog", *HESC*, https://hesc.co.za/species-hesc/african-wild-dog/ (accessed 6 January 2020).

4.6   Geographic Information Toolsets

The previous sections have discussed using many different sources of satellite images or other data but the process and requirements behind these have not been clearly explained. This section will examine and compare the suitable data providers and tools to use said data in the context of African development. It will also compare these to the costs of launching independent satellites as well as the benefits and downsides of doing this.

With the amount of satellite data available either freely or at low cost, there is very little reason for countries, particularly African countries that have smaller budgets than those of other parts in the world, to create their own Earth observation satellites. If they do, it should be primarily for the use of capacity-building and education rather than for practical use at this stage. If integrity of the data from an existing data source is deemed to be compromised, there are enough providers so that a similar dataset may be acquired for far less than the cost of a satellite and launch.

4.6.1   Geographic Information Systems (GIS) Definition

GIS is the application of different layers of useable data to some sort of map. For example, a base layer could be roads and a useful data layer placed on top of that could be the location of petrol stations on roads. Most modern GPS systems such as mapping applications on smartphones etc. are a form of GIS.

GIS has existed since before satellite imagery was available. Initially information was simply drawn onto layers of a hand drawn map, derived using old cartography techniques. In some areas, particularly those of high population density, these proved highly accurate, as far back as Ancient Greece (Fig. 4.24).

However, true widespread accuracy was only gained after the invention of flight, when planes could take photos of large spans of areas previously impenetrable by land or sea. Many world maps are currently still based on aerial photography. Aerial photography can still have significantly higher spatial resolution than most satellites, at the downside of temporal resolution. Multiple flights are expensive, and it may take months just to cover a few thousand kilometres worth of images.

Here is where satellite imagery has the advantage. Modern providers can provide repeat images of areas within half-a-day, which can be very useful in highlighting rapid changes associated with the developing world. Most modern map systems such as Google Maps rely primarily on high-res satellite imagery. Google Maps and Google Earth and other competitors themselves are basic GIS systems useable by anyone with a computer or a smartphone. Road and other layers can easily be added or removed, and as noted can themselves be useful in some applications in development for Africa.

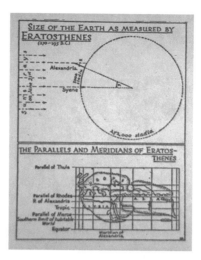

Fig. 4.24 Ancient Greek map showing lines of latitude and longitude ("Eratosthenes Measurement of the Earth", *Henry Davis Consulting*, 2003)

4.6.2 Comparison of Two Common GIS Software Packages

The two most notable desktop software packages for GIS are QGIS[64] and ArcGIS.[65] Figure 4.25 shows the QGIS interface. ArcGIS's interface is quite similar. There are a number of similarities and differences between the two, but the primary difference is that QGIS is open-source and free and ArcGIS costs US $100 a year for the most basic features.[66] This immediately puts QGIS at an advantage for governmental organisations in Africa, which have very small budgets to begin with.

As QGIS is open-source, it also accepts, or can be made to accept, far more varied types of data. Versions of it can run on any operating system, while ArcGIS is exclusively for Windows. Both accept scripting with Python or other languages, but QGIS is far more open to added code. ArcGIS has more support due it being a paid-product and has far more built-in spatial analysis features than QGIS. These can be added to QGIS, however. ArcGIS has more ease-of-use as well, and a simpler interface. Both allow online base-maps such as Google Maps, but ArcGIS has far more of these sources available. Both allow a large variety of plugins, some of which for ArcGIS cost money but are also highly-specialised.[67]

---

[64]"QGIS", qgis.org (accessed 6 January 2020).

[65]"ArcGIS", www.arcgis.com (accessed 6 January 2020).

[66]Ibid.

[67]"27 Differences Between ArcGIS and QGIS—The Most Epic GIS Software Battle in GIS History", *GISGeography*, 8 August 2015, https://gisgeography.com/qgis-arcgis-differences/ (accessed 6 January 2020).

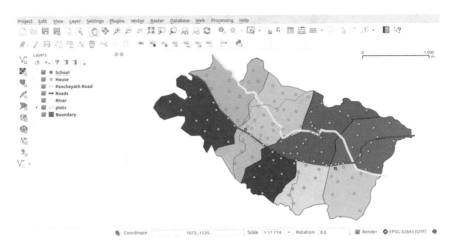

Fig. 4.25 QGIS interface (Path slopu [CC BY-SA (https://creativecommons.org/licenses/by-sa/4.0)], "QGIS 2.8 screenshot", 12 October 2018, https://commons.wikimedia.org/wiki/File:QGIS_2.8_screenshot.png (accessed 6 January 2020))

Overall QGIS can be thought of as a powerful, free, raw data processor, while ArcGIS is easier to use and may be better tailored to specific applications.[68]

The trade-off then comes between training time of technicians to use QGIS and the cost of ArcGIS. In terms of the African setting, enough basic data processing is available in QGIS with minimal required training that it should be a software of choice for almost all African countries.

4.6.3 ESA Snap

ESA's SNAP[69] software should be mentioned here. It is an open-source software developed specifically to deal with data from the Sentinel satellites. While it does have many GIS layering capabilities it should be noted that this software is more for the processing of satellite images, which may then be imported into a GIS software such as QGIS. GIS data can also be imported into SNAP, depending on the needs of the user.

SNAP is powerful in that it allows quick processing of both multi-band optical imagery and advanced processing of SAR imagery. Since Sentinel data and the SNAP software are both free and integrate extremely well together, it is a very powerful tool that African countries may use. A prime example of the use of Sentinel data can be to identify illegal fishing or pirate vessels. This can be done in a matter of minutes by anyone with some basic training and decent computer skills (Fig. 4.26).

---

[68]Ibid.

[69]SNAP Sentinel Processing Toolbox, https://step.esa.int/main/ (accessed 6 January 2020).

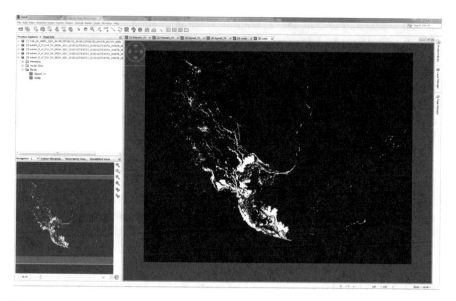

Fig. 4.26 Flood mapping using sentinel SAR data and SNAP ("Recommended practice flood mapping", *UN-SPIDER*, https://www.un-spider.org/book/export/html/10313 (accessed 6 January 2020))

4.6.4 Adoption of GIS Technologies in Africa

Unfortunately, very few African governments are aware of the power of tools such as these to solve potential governance issues such as water management etc. Only those that are more developed such as South Africa or Nigeria have good use of such tools, which interestingly are some of the countries that have a Space Policy. Some other countries do use them but not nearly to their full potential.

If widespread use of such tools in Africa could be achieved, the benefits would be immeasurable as described in all the areas earlier. However, due to Africa's tough and large landscape, having representatives reach every region that needs these tools and spend enough time to educate on how to use them is highly challenging. Even though these GIS tools are used for projects in Africa, it is usually by external companies for construction projects. African governmental organisations for the most part are not using or understanding the full capability of such tools themselves.

A possible avenue is to approach national governments with the capabilities of such software and emphasise the cost saving that such technologies could provide. QGIS is free, can use up-to-date and free satellite images, and can process extremely quickly. Provided an organisation that deals with some form of governance has at least one available PC and some sort of Internet connection, there is no reason that a person should not be trained in this powerful technology, whether it be identifying areas that may be very prone to floods, or tracking mosquitos that may possibly carry malaria. Newly-trained personnel may also be overwhelmed by the availability of data and processing techniques. Therefore, basic guidelines should be established to allow technicians to keep their focus on important areas and datasets, such as catchment around one main dam supplying a city.

Some organisations are focussed on training in GIS initiatives. RCMRD[70] is an inter-governmental organisation based in Kenya supported by 20 African states that develops capacity in the mapping field. However, this organisation, and other such centres such as UNOOSA-backed ARCSSTE-E and ARCSSTE-F in Morocco and Nigeria respectively, require more awareness and more use by African states.

These objectives will still take significant time and money to achieve, but the benefits far outweigh these issues, and the costs of a disaster are insurmountable compared to that of tools that can easily help prevent it. Furthermore, as machine learning improves, these tools will become more and more automated and much simpler for governments to implement. This should occur in the next decade or two, and will hopefully see significant benefit to governance in Africa.

4.7 European Space Agency's Involvement in Supporting African Development

The European Space Agency has invested heavily in projects to support world development, particularly in Africa in relation to the Sustainable Development Goals as well as other objectives. This section will provide an overview of the technology in use for these projects as well as the projected outcomes and benefits. Some of these have been noted in previous sections, but more background will be given here.

4.7.1 Copernicus

Copernicus has been mentioned already, as the overall programme involving the Sentinel satellites as well as their operation, data acquisition and use. It is a partnership between the European Commission, ESA and EUMETSAT. The programme was specifically designed to provide high spatial and temporal resolution imagery of the entire Earth, and to be easily available for both environmental and humanitarian purposes. Copernicus was previously known as Global Monitoring for Environment and Security (GMES).

Copernicus is divided into three segments:

- Space segment
- Ground segment
- Service segment.

---

[70]"RCMRD", https://www.rcmrd.org/ (accessed 6 January 2020).

Table 4.3 Satellites of the Copernicus programme

Sentinel	Number Operational	Number Planned	Description
1	2	4	SAR Imagery for all-weather applications. C-band at 5 × 20 m standard resolution
2	2	4	Multi-spectral optical comprising of visible and near-infrared bands. 10 × 10 m resolution
3	2	4	Ocean, land and climate monitoring. 4 instruments to monitor ocean and land colour and temperature, altitude and water vapour
4	0	2	Instrument payload on EUMETSAT MTG-S geostationary satellites. For air quality monitoring applications. Positioned over Europe with some available African coverage
5P	1	1	Precursor satellite to Sentinel-5
5	0	2	Instrument payload on MetOp-S polar orbiting satellite. Air quality measurements in tandem with Sentinel-4
6	0	2	Follow on from JASON satellites. Sea altitude measurement

The space segment obviously involves all the satellites necessary to provide this data. Table 4.3 provides an overview of the main satellites and instruments planned and in-use for the Copernicus Programme.

It should be noted that many existing and other planned missions complement these main missions, such as the European missions SPOT, MSG, MetOp, TerraSAR-X, COSMO-Skymed, DMC, PLEIADES, as well as non-European providers including LANDSAT, DigitalGlobe, GOSAT and RADARSAT.

The ground segment of the Copernicus programme is a mix of existing and new ground station infrastructure between ESA and its partners. The service segment involves all the IT infrastructure necessary to process and provide the data received from the space segment through the ground segment. Furthermore, Copernicus is complemented by in-situ measurements such as atmosphere and water quality, co-ordinated by the European Environment Agency. Figure 4.27 shows the entire hierarchy of the programme.

The Copernicus Programme was not centred on European needs but rather global needs and, from its inception, was designed to aid Africa. Copernicus has been successfully implemented and has numerous specific examples of providing support to different humanitarian situations, as varied as detecting algal blooms to rescuing lost explorers.[71]

---

[71]"Copernicus Sentinel-1 to the Rescue", *ESA*, 13 December 2019, https://sentinel.esa.int/web/sentinel/news/success-stories/-/asset_publisher/3H6l2SEVD9Fc/content/copernicus-sentinel-1-to-the-rescue (accessed 10 February 2020).

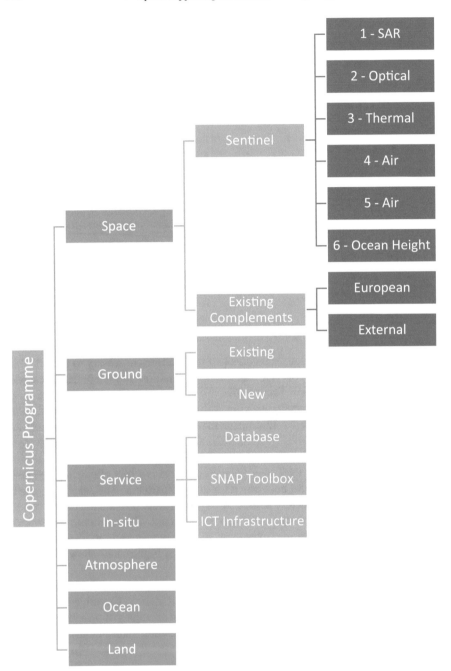

Fig. 4.27 Organisation of Copernicus

Most of the space applications applicable to Africa's developmental needs can be provided by Copernicus alone, at no cost (apart from the IT infrastructure needed to access the data) and with minimal training needed. Copernicus is a prime asset for African governments, organisations and even individuals, in support of development. There is no other single programme that provides such a large variety of high-quality data for the challenges Africa faces. Yet again, as noted above, the difficulty occurs in education and acceptance of the available data and the use thereof. It is the responsibility of those in Africa who understand the capability and possibilities of such technology to share and educate so that programmes such as Copernicus get used to their full extent.

4.7.2 SWAY4Edu2

SWAY4Edu2 is a Satcom-based education project created by ESA.[72] It is a pilot project to determine the possibility of high-quality distance education over satellite communication. The first phase involved the use of high-broadband satellite video to educate a group of students at a rural school in South Africa and a rural school in Italy at the same time. The project also provided 20 Satcom radios to extremely rural areas of the D. R. Congo to enable distance education. Figure 4.28 shows the relay setup for distance learning.

This project, while not extensive, showed that modern technology could provide remote education in areas with extremely little existing infrastructure. It proves that if increased in scale, which modern satellite technologies should easily allow, remote education via satellite is feasible.

4.7.3 GlobWetland Africa

Africa is home to some of the most expansive and unique wetland environments in the world, hosting thousands of diverse species and fragile ecosystems. They also indirectly bolster the water cycle and drainage systems that support a significant amount of African agriculture. However, they do not appear as directly useful to humans, which means they are in severe danger.

Most frequently drained for agriculture or construction, wetlands can disappear very quickly. To combat this, ESA has partnered with the Ramsar Secretariat to create GlobWetland Africa (GW-A), an extension of the GlobWetland 1 and 2 projects.[73] This is in conjunction with the Ramsar Convention on Wetlands of International Importance, which 50 countries have signed. This programme provides significant benefits as, without this satellite data, there would be very little information about the

---

[72]"ESA Projects", https://business.esa.int/projects/sway4edu2 (accessed 6 January 2020).
[73]"GlobWetland-Africa", https://globwetland-africa.org/ (accessed 10 January 2020).

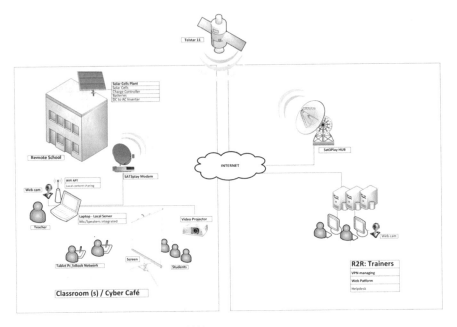

Fig. 4.28 SWAY4Edu infrastructure (Ibid.)

state of wetlands in Africa. The programme thus prevents wetlands from disappearing without notice.[74]

GW-A mainly uses Sentinel data with some Landsat images to provide detailed and up-to-date information about the condition of all African wetlands. It then provides this data to African organisations so they can make informed decisions about the protection of wetlands.

An open-source GW-A toolbox has been developed, which works with QGIS, another open-source platform, making accessing and manipulation of data completely free. The products provided by GW-A are shown in Table 4.4.

Figure 4.29 shows an example of the wetland inventory product of Lake Victoria in Uganda. The darker the colour, the higher the probability of the area being a

[74] For additional information see: M.O. Rasmussen, R. Guzinski, C. Tøttrup, M. Riffler, M. Paginini, and B. Koetz, "Earth Observation for Water Resource Management and Sustainable Development", in *Earth Observations and Geospatial Science in Service of Sustainable Development Goals: 12th International Conference of the African Association of Remote Sensing of the Environment,* ed. S. Wade (Cham: Springer, 2019), 3–14; M.M. El-Hattab, A. Ahmed, and M. El-Raey, "Morphometric Analyses of Tarhuna Drainage Basins to Accesses Groundwater Potential Using GIS Techniques", in *Earth Observations and Geospatial Science in Service of Sustainable Development Goals: 12th International Conference of the African Association of Remote Sensing of the Environment,* ed. S. Wade (Cham: Springer, 2019), 57–68; and K.B. Mfundisi, "Spatiotemporal Analysis of Sitatunga (Tragelaphus Spekei) Population's Response to Flood Variability in Northern Botswana Wetlands: Implications for Climate Change Mitigation", in *Earth Observations and Geospatial Science in Service of Sustainable Development Goals: 12th International Conference of the African Association of Remote Sensing of the Environment,* ed. S. Wade (Cham: Springer, 2019), 103–116.

Table 4.4 GlobWetland Africa products

Product	Description
Wetland inventory	Identification, mapping and delineation of wetlands
Wetland habitat maps	Land use in wetland areas for threat derivation
Water cycle regime	Historical variation of rainfall and water table levels
Water quality parameters	Cyanobacteria, algal and sediment levels
River basin hydrology	Regional catchment water information such as soil moisture and surface water extent
Mangroves mapping	Creating an inventory of mangroves and identifying spatial distribution

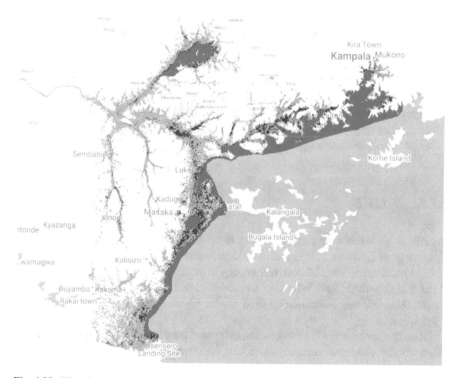

Fig. 4.29 Pilot site on Lake Victoria in Uganda—Wetland Inventory (Ibid.)

wetland. This data is solely derived from SAR interpolations and is automatically generated, making it easy to detect wetlands throughout Africa.

In order to assist most African countries to comply with the Ramsar convention, as well as reach their SDGs, GW-A has further implemented cloud-based processing to allow the obtaining of data with lower-bandwidth Internet connections.

GW-A is a powerful set of products and toolboxes that, if used correctly, will help keep many wetlands from disappearing and harming both a large amount of wildlife as well as fragile hydrological systems.

4.7.4 EDRS Global

EDRS (European Data Relay System) Global, formerly known as GlobeNet, is a partnership between ESA and Airbus Defence and Space to provide an ultra-high bandwidth relay system from LEO satellites to the GEO backbone and then back to Earth.[75]

The benefit of this relay system is that it can provide constant coverage from LEO satellites, i.e. there will not be a limited availability window. Because of this, it can be used in real-time situations where fast action is key, such as emergency scenarios. The EDRS satellites are equipped with advanced laser communication terminals (Fig. 4.30), allowing a bandwidth of up to 1.8 Gbps.

Two of these satellites are operational already, with a third planned to increase throughput. It should be noted that their orbit allows them to serve Africa extremely effectively, as they are located directly over the continent. Also, EDRS can be used by aerial drones for constant security monitoring applications.

4.7.5 B-LiFE

B-LiFE is an interesting project arising from collaboration between ESA and the University of Louvain.[76] It is a mobile deployable lab designed to be able to help combat pandemics using Satcom technology to be able to analyse results much faster. It has a highly-noticeable inflatable antenna that allows high-bandwidth communication. The lab was successfully used in the Ebola outbreak in Guinea in 2015 (Fig. 4.31).

It has further benefits in emergencies as it provides a back-up communication link when normal cellular communications may be overused.

---

[75]"ESA Projects", https://www.esa.int/Applications/Telecommunications_Integrated_Applica tions/EDRS/EDRS_Global (accessed 10 January 2020).

[76]"ESA Projects", https://business.esa.int/news/b-life-%E2%80%93-life-saving-labs-lightning-speed (accessed 10 January 2020).

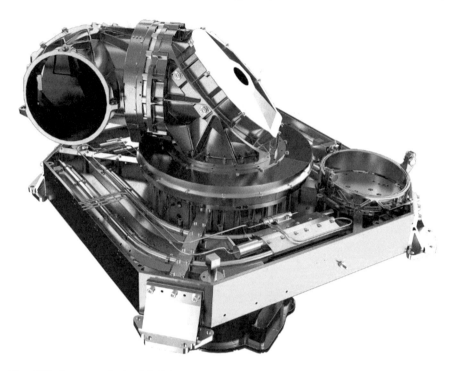

Fig. 4.30 Laser terminal of EDRS Satellite (DLR, CC-BY 3.0/CC BY 3.0 DE (https://cre
ativecommons.org/licenses/by/3.0/de/deed.en), "EDRS Optical Terminal", 2015, https://com
mons.wikimedia.org/wiki/File:European_Data_Relay_System_laser_communication_terminal.
jpg, (accessed 15 February 2020))

4.7.6   SatFinAfrica

SatFinAfrica was a pilot ESA programme that successfully developed into a commercial business to the benefit of many African citizens.[77] It involved a relatively simple
premise, to create ATMs connected by satellite rather than traditional communication infrastructure, to enable those in remote areas to transfer or withdraw money.
It is especially useful in Africa, where many people may work far from where their
family stay and need to send money to them without having to transport cash. The
current platform uses high-data rate off-the-shelf antennas for low latency and cost.

The higher supply of money in a region due to more banking options further aids
development by making local and micro-economies more liquid, enabling people to
pay for products and services easier and small businesses to grow alongside emerging
financial services.

---

[77]ESA, "SatFinAfrica—Reliable and secured financial services in remote areas", https://business.
esa.int/projects/satfinafrica (accessed 10 January 2020).

Fig. 4.31 B-LiFE deployed (B-LiFE, https://business.esa.int/news/b-life-%E2%80%93-life-sav
ing-labs-lightning-speed (accessed 10 January 2020))

4.7.7 ESA TIGER

ESA TIGER[78] is an initiative started in 2002 with the aim of protecting water supplies
in Africa through capacity building and the applications of satellite imagery. The
programme has had a large number of projects targeting different developments as
well as educational courses in the field of earth observation for water. All facets of
water have been examined such as catchment areas, water quality, ground and surface
water as well as various outreach projects.

The project has reached numerous goals and has built large capacity both in the
research and governmental bodies of Africa. It has helped form a solid basis for
African organisations/governments to take available data such as that from Coper-

---

[78]ESA, "Tiger initiative", https://www.tiger.esa.int/ (accessed 10 January 2020).

nicus and use it to better African water management with capacity for continued development. The continued easier availability of satellite data means that water management in Africa should improve thanks to the basis built by TIGER.

4.8 A Modelling of the Sustainable Development Goals with Their Risks and Available Support from Space

This section discusses the SDGs (as covered in Chap. 1) using soft mathematics rather than policy and organisational discussion. It will examine the effect of certain SDGs on others and the risks associated with each, as well as how space may benefit both.

While there are some fundamental differences between each SDG, for ease of modelling they have been grouped into the categories shown in Table 4.5. Furthermore, each SDG has been given a monetary, time, and capacity (education and person-effort) cost needed weighted from 1 to 5, for African needs. Note that this analysis is highly-simplified and is intended to present an easy-to-understand viewpoint of Africa's needs with regards to the SDGs.

Based on the above costings, an average cost per SDG per category has been calculated, shown in Table 4.6.

It can be seen that reaching Africa's health and economic targets are the most challenging goals. However, all the SDGs have an effect on each other, which means that some will lower the cost of others. For example, education is a fundamental support structure for all the other goals, leading to far easier achievement of all other goals (Fig. 4.32).

In the same way, education is mainly supported by economic sustainability (Fig. 4.33).

By using the above principles, a matrix was created showing how much each goal category supported the others based upon the same rating scale of 1–5, as seen in Table 4.7. Again, these ratings were calculated given the evidence presented in this publication.

Finally, the extent to which space applications and technology can help in the achievement of each goal category was also rated from 1 to 5 and can be seen in Table 4.8. These ratings have been based on the costs and practicality of space applications in support these goals, as discussed in the previous sections.

These combined sections of data can then be visualised as in Fig. 4.34. Combining each section can give some sense of the importance of each goal category for African needs and allows a ranking to be determined for possible investments in space technologies and programmes to support them. This is shown in Fig. 4.35.

Note that this result is based on some qualitative measurements. However, the rankings have been determined by actual African needs and historical evidence of necessary requirements to achieve similar goals. It can be seen that major investments

Table 4.5 Categorisation and weighted costs of SDGs

SDG	Monetary cost	Time cost	Capacity cost	Category	Reasoning for cost
1. No poverty	4	4	4	Economic sustainability	Development of wealth requires initial investment and education for fostering of investment
2. Zero hunger	5	5	5	Health	With an ever-growing population, Africa suffers from mass starvation, the fixing of which requires effort on all fronts
3. Good health and well-being	3	3	5	Health	Health relies on education above all
4. Quality education	3	5	4	Education and capacity building	Education requires education and infrastructure, with developed generational experience that requires time
5. Gender equality	1	1	2	Human rights, fairness and safety	Gender equality should be easy to achieve, provided governments have better policies and enforcement
6. Clean water and sanitation	5	4	5	Health	Water supplies are facing ever increasing strain throughout Africa
7. Affordable and clean energy	3	4	5	Economic sustainability	Clean energy is almost as affordable as non-renewables, especially given Africa's natural resources. However, development on African terrain requires huge commitments

(continued)

Table 4.5 (continued)

SDG	Monetary cost	Time cost	Capacity cost	Category	Reasoning for cost
8. Decent work and economic growth	4	5	3	Economic sustainability	Once basic infrastructure has been developed, decent work can be provided, but this takes time
9. Industry, innovation and infrastructure	5	3	4	Economic sustainability	Modern techniques enable infrastructure to be developed quickly, however large economic costs are required
10. Reducing inequality	4	5	2	Human rights, fairness and safety	Reducing inequality requires the other SDGs to be met first, in order to raise standards-of-living
11. Sustainable cities and communities	5	3	3	Environmental sustainability	Same reasoning as SDG 9
12. Responsible consumption and production	2	1	5	Environmental sustainability	With modern technology, it is only more costly by a small margin to be responsible in consumption and production. However, vast education is needed to achieve this
13. Climate action	3	3	4	Environmental sustainability	Since much of Africa's infrastructure still needs to be developed, this can be done environmentally friendly from the beginning, which will reduce costs

(continued)

Table 4.5 (continued)

SDG	Monetary cost	Time cost	Capacity cost	Category	Reasoning for cost
14. Life below water	2	1	4	Environmental sustainability	Both SDG 14 and 15 require effort and education more than time or money. Sadly, they need to be given more attention by African governments
15. Life on land	2	1	4	Environmental sustainability	See above
16. Peace, justice and strong institutions	3	2	5	Human rights, fairness and safety	This can be done quickly and without too much cost provided there is effort and dedication by African governments
17. Partnership for the goals	1	1	3	Education and capacity building	Co-operation is relatively easy provided bodies understand the benefits it offers

Table 4.6 Weighted cost (1–5) per SDG category

Category	Average monetary cost	Average time cost	Average capacity cost	Average cost
Economic sustainability	4	4	4	**4**
Health	4.33	4	5	**4.44**
Education and capacity building	2	3	3.5	**2.83**
Human rights, fairness and safety	2.67	2.67	3	**2.78**
Environmental sustainability	3.4	2.2	4	**3.2**

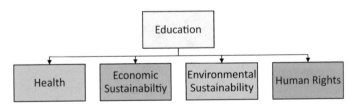

Fig. 4.32 Education boosting other goals

Fig. 4.33 Economic sustainability supporting education

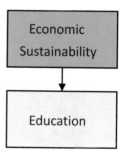

Table 4.7 Goal categories supporting other goals (weighted 1–5)

		Goals affected					Average score
		Economic sustainability	Health	Education and capacity building	Human rights, fairness and safety	Environmental sustainability	
Goals affecting	Economic sustainability	–	4	4	2	2	**3**
	Health	3	–	2	1	2	**2**
	Education and capacity building	5	5	–	5	5	**5**
	Human rights, fairness and safety	1	3	3	–	3	**2.5**
	Environmental Sustainability	4	5	3	3	–	**3.75**

Table 4.8 Ability of space applications to support goal categories, rated 1–5

Category	Ability of space to support each goal
Economic sustainability	2
Health	3
Education and capacity building	4
Human rights, fairness and safety	2
Environmental sustainability	5

in space technology to support African development should be done primarily for environmental protection and education. Through programmes such as Copernicus, this has already started.

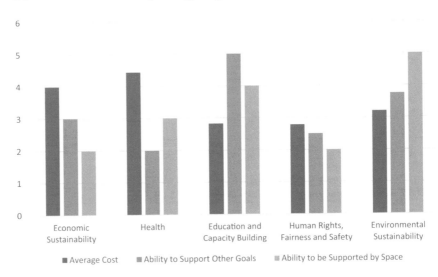

Fig. 4.34   Rating of goals in terms of costs and abilities

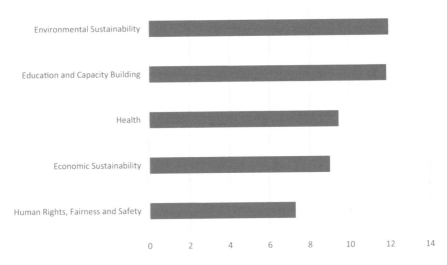

Fig. 4.35   Generalised ranking of goal categories

4.9   Conclusion

Space applications and satellite data can aid governance in a large variety of ways. In terms of defence, both visible light imagery and SAR can be used to spot insurgent activity such as camps and vehicles, with SAR even allowing this through vegetation and cloud cover and at night. This may be extremely useful for tracking African terrorist activity such as Boko Haram etc. New defence satellites are also mostly not-necessary, with even free data sources providing enough information to track potential threats.

In terms of health governance, satellite images may be used to directly track conditions favourable to disease vectors such as mosquitos. As such they can determine when a wave of a disease such as malaria may be approaching and enable health authorities ample preparation. Furthermore, telemedicine can be done over space-based communication, allowing a surgeon from another region of the world to operate on an African patient.

Space technology can also provide vital help in the management of water. Smaller catchment regions to larger hydrological cycles may be observed and modelled, enabling predictions of droughts as well as the formation of preventative measures. Through good water governance, food security can be ensured.

Africa contains a massive amount of biodiversity that sadly many governments are unable to manage effectively due to cost constraints and large natural areas. Satellites enable individual animals to be tracked, as well as habitats to be observed for destruction, meaning that governments can remotely keep track on the well-being of animals, and find problem areas quickly without the need for field work.

To facilitate the above space application areas, a number of toolsets exist to easily work with the data. QGIS is free and quite powerful. SNAP, developed by ESA, complements the freely-available Sentinel data and allows extremely fast processing to occur, even of the most complex SAR images. ESA's Copernicus programme, one of many that ESA has developed, provides a resource of high spatial and temporal resolution satellite imagery encompassing many types. If used correctly by African governments, this and the other programmes may prove extremely beneficial to African governance, and already have in some ways. African governments mainly need the capacity to use these tools and data sources. Fortunately, suppliers such as ESA are working on educating leaders about their potential.

Finally, through a basic qualitative approach, the 17 SDGs as mentioned in Chap. 1 were analysed, and it was found that space applications can be most useful for education and environmental sustainability in Africa, using the techniques discussed previously.

Printed in the United States
by Baker & Taylor Publisher Services